BRITISH JET AIRCRAFT

Adrian Vicary

PSL Patrick Stephens, Cambridge

First published in 1982

British Library Cataloguing in Publication Data

Vicary, Adrian
British jet aircraft.
1. Jet planes—History
I. Title
629.133'349 TL670

ISBN 0-85059-589-4

Photoset in 10 on 10 pt Baskerville by Manuset Limited, Baldock, Herts. Printed in Great Britain on 115 gsm Fineblade coated cartridge, and bound, by The Garden City Press, Letchworth, Herts, for the publishers, Patrick Stephens Limited, Bar Hill, Cambridge, CB3 8EL, England.

Front endpaper *An 11 Squadron pilot watches while ground-crew start the engines of his Lightning F Mk 6* (MoD).

Rear endpaper *Sepecat Jaguar GR 1 with a full load of rockets, bombs and drop tank* (MoD).

Frontispiece *Unusual view of a Lightning at RAF Akrotiri, with air brakes and cockpit open* (RAF).

Contents

Preface

In the following pages lies what I hope will be a stimulating and interesting survey of the development of the British jet-powered aeroplane. Since the appearance of the Whittle-powered Gloster machine of the early 1940s, Britain's aircraft industry, despite continuous economic and other difficulties in post-war years, has produced as wide a range of types as any other leading nation, from pure research aircraft through jet fighters, bombers and trainers to airliners, a supersonic transport and even a successful executive jet.

In the last 15 years or so the massive expenditure required for each new development has necessitated collaboration with other countries to produce the complete aeroplane. Despite difficulties arising through political differences some fine results have been achieved, notably the Jaguar and Tornado.

The criterion for types to be included in the book is that each should be a pure jet-powered aeroplane. Therefore the turbo-prop and gas turbine-powered helicopter will not be found in its pages. Also omitted are the engine test-beds (like the Lancaster series) which, although performing invaluable work in trying new engines subsequently used in many of the aircraft included, did not rely entirely on the turbojets to carry out their flights.

Each type is included in chronological order of date of first flight, this plan being adopted to maintain the history accent which is the main theme. The background descriptions have been deliberately shorn of the majority of technical detail for two reasons: firstly to make them more readable; and secondly to allow the relevant data to be given in tabulated form, the layout of which has been chosen to give a quick and clear comparison between types.

Right *Camouflaged Jet Provost Mark 4 in the Forward Air Control role at RAF Brawdy with the Meteor F8 target tug VZ467 'Winston'* (Richard Gardner).

Introduction

The first flight of the Gloster E.28/39 on May 15 1941 was the culmination of some 11 years' work by Frank Whittle (now Air Commodore Sir Frank Whittle, RAF (ret'd)) on the then-revolutionary jet engine. The first patent had been applied for in January 1930 and his ideas were brought to the attention of the Air Ministry. Unfortunately they were turned down on the grounds of the cost involved in developing such a new and untried concept. He then turned to private industry to bring his ideas to fruition but again met with a lack of success.

As an officer in the Royal Air Force his career in that service progressed and by the end of 1934 he had completed an Officer's Course of Engineering at RAF Henlow, whence he proceeded to Cambridge. In July 1937 he graduated with First Class Honours in the Mechanical Engineering Tripos and was promoted to Squadron Leader. The period at Cambridge was to prove crucial in the development history of the jet engine, for it was there that two former RAF officers, Tinling and Williams, approached Whittle with a proposal to enable his ideas to achieve practical form. In March 1936, with financial assistance from a firm of investment bankers, Power Jets Ltd was formed. Design of the first engine was commenced in June and it was constructed with the help of the British Thomson-Houston Company Ltd, of Rugby. It made its first bench test on April 12 1937 and testing was continued in a disused foundry at Lutterworth.

Now that the engine seemed to be a practical proposition the Air Ministry stepped in and provided the necessary financial backing to develop an engine suitable for flying, the contract being placed with Power Jets on July 12 1939. They also ordered, from the Gloster Aircraft Company, the airframe in which the engine was to be installed.

It may be said in truth that the birth of the jet-propelled aeroplane was as fraught with difficulty as was that of the heavier-than-air flying machine at the outset of the 20th century. However, the principle of jet propulsion had been demonstrated as long ago as the 1st century AD, when the Greek philosopher, Hero of Alexandria, had built a steam-powered model. This consisted of a pivoted hollow metal sphere spun by steam-emitting nozzles around its circumference. In 1913 the French engineer, René Lorin, published patents which propounded the basic principles of jet propulsion using the exhaust gases from an internal combustion engine.

In Britain, at the Royal Aircraft Establishment, Farnborough, a team of scientists under the leadership of Dr A.A. Griffith began, in 1925, to research into gas turbines as a source of power for aircraft. However, their work was concentrated on transmitting the power via a propeller—the turbo-prop—whereas Whittle concentrated on the pure jet, using the thrust of the stream of hot exhaust gases for propulsion. Although Whittle's patent, taken out in 1930, was the first to cover a practical engine, the long period between that action and the successful bench testing of the engine allowed others to get into the air first.

In Germany an engine designed and built by Dr Hans-Joachim Pabst von Ohain in 1935 aroused the interest of Ernst Heinkel, who secured Ohain's services in April 1936; with the result that an aircraft, the He 178, powered by a developed version of the engine, first flew on August 27 1939. And in Italy Ing Secondo Campini designed an aeroplane and its engine which was built by the Caproni concern and which made its first flight on August 28 1940. It later made a long cross-country flight from Milan to Rome on November 30 1941.

The Germans wasted no time in developing the turbojet further, several streams of research being explored. The Heinkel company followed their He 178 with a completely new design planned from the outset as a fighter to be powered by two engines developed by von Ohain from his earlier powerplant. Although the airframe of the first prototype He 280, as it was called, was ready some months before the engines were available, it was still able to get into the air before the Gloster E.28/39 on April 2 1941.

So the Gloster E.28/39 was the world's fourth jet-engined aeroplane to take to the air.

The Aircraft

Gloster E.28/39

The Pioneer, as it was appropriately if unofficially dubbed, was designed by George Carter, chief designer of the Gloster Aircraft Company, to the Air Ministry specification E.28/39 calling for an experimental aeroplane to flight test the Whittle jet engine. It was to be capable of carrying an armament of four 0.303-in Browning machine-guns together with their ammunition, although it would not be required to do so during the initial trials. In the event the guns were never fitted.

Before detailed design was begun, the layout of the airframe with regard to the installation of the jet-pipe had to be decided. With a short jet-pipe and a tail unit mounted on twin booms a better performance could be achieved, but at that time the effect the hot gases would have on the airflow over the tailplane was unknown. Therefore the alternative of carrying the jet-pipe through the fuselage and exhausting below and behind the tail was chosen.

Two aircraft were to be built and construction was started in the experimental department at Hucclecote. However, Gloster's main work at the time, mid-1940, was the production of Hurricane fighters and this meant that the factory was likely to receive the unwelcome attention of the Luftwaffe. Accordingly the first airframe was moved to one of the company's dispersal premises in Cheltenham, Regent Motors' garage. *W4041* was completed there and in the first week of April 1941 was taken by road back to Hucclecote for initial testing. The company chief test pilot, Flight Lieutenant P.E.G. Sayer, made taxiing trials powered by the 'non-flying' W.1X engine and during these the aircraft briefly became airborne. Frank Whittle was in attendance and also took the controls in these first trials. After further ground testing and modification of the nose landing gear, the aircraft was dismantled and taken to Cranwell for flight trials. The airworthy Whittle W.1 engine was fitted.

In the evening of May 15 1941 the aircraft took

The Gloster Whittle E.28/39 photographed at Cranwell on May 15 1941 (Gloster photo via Mike Bowyer).

off and completed its first flight of 17 minutes' duration, in the hands of Flight Lieutenant Sayer. The performance of both airframe and engine was thereafter extensively explored and modifications applied in the light of experience gained. Michael Daunt took over as chief test pilot after Sayer had been killed flying a Typhoon off the Northumberland coast in October 1942.

The second aircraft, *W4046*, was completed and, fitted with the Rover W.2B, made its first flight on March 1 1943 piloted by John Grierson, Gloster's development test pilot. He made the first cross-country flight in the same aircraft on April 17, when it flew from Edgehill to Hatfield to give a demonstration for Winston Churchill.

Both machines spent some time at the RAE Farnborough, where they were flown by service pilots. During this period, on July 30 1943, *W4046* was destroyed. At high altitude the grease on the aileron controls solidified, causing them to jam and the aircraft went into an inverted spin. The pilot, Squadron Leader Douglas Davie, was thrown out of the cockpit but was able to make a parachute descent, thus becoming the first pilot to escape from a jet aircraft. The first aircraft continued to fly until it was dismantled at Farnborough early in 1945. In April 1946 it was re-assembled and put on display in the Science Museum, Kensington.

Gloster Meteor

Any hope that an operational fighter could be developed from the E.28/39 was soon dispelled when it was realised that the evolution of an engine with the thrust necessary to power an aircraft carrying armament and armour protection for its pilot was going to be a slow process. Therefore a twin engine installation was required, resulting in the design that was to emerge as the Meteor. The Air Ministry specification F.9/40, written around the designs previously submitted by George Carter for such a machine, was issued to Gloster in November 1940 and the design was approved for prototype construction in February 1941. Initially the order was for 12 airframes in which engine and other equipment trials would be carried out, but this total was later reduced to eight.

The Rover W.2B engine, which had powered the second E.28/39, was planned to power the prototypes but difficulties with the performance of the engines led to delays in production and this in turn resulted in the airframes being ready before powerplants were available for them. Meanwhile, at de Havilland, the development of another jet engine, designed by Major F.B. Halford, had been in progress. This was in a more advanced state of readiness and with a pair fitted to the fifth prototype, *DG206/G*, the first flight was made at Cranwell by Michael Daunt on March 5 1943. Airworthy Rover W.2B/23 engines became available in time for the first prototype, *DG202/G*, to make its maiden flight on July 24, and the third aeroplane to fly was *DG204/G* fitted with Metropolitan-Vickers F.2s on November 13. The latter was of significance in being the first occasion on which a British axial-flow engine had taken to the air, all previous flights being made with the original Whittle-type centrifugal-flow engines.

Following on from the F.9/40s, 20 Meteor F Mk 1s were ordered. They were to receive Welland engines developed from the W.2B/23 by Rolls-Royce, who had taken over production of the Whittle engines from Rover. *EE210/G* was the first to fly on January 12 1944, the honours going again to Michael Daunt. By July enough aircraft were ready to make initial deliveries to an operational unit, No 616 Squadron based at Culmhead, Devon. The type's stay there was shortlived. The squadron

The RAF's last operational Gloster Meteor, F Mk 8 VZ467, *photographed at RAF Brawdy in 1976; nicknamed 'Winston', this aircraft was then serving as a target tug* (MoD).

moved before the month was out to Manston, Kent, to help combat the new flying bomb menace. Despite frustrating problems with the guns, a Meteor made a successful attack on a V1 by tipping it over with its wingtip. This occurred on August 4 and the pilot involved was Flying Officer Dean. The score of flying bombs destroyed by the jets soon mounted as the gun problems were cleared up.

The Meteor Mk II, which should have appeared next, was cancelled owing to the proposed use in it of the de Havilland Goblin engine (a development of the H.1 used to power the first flying prototype), all production of which was required for de Havilland's own jet fighter, the Vampire. So the next mark to be built was the F Mk III, initially fitted with Welland engines but from the 16th airframe onwards with the more powerful Rolls-Royce Derwent. The Mk IIIs soon helped to bolster the small numbers of Mk Is in service, firstly joining No 616 Squadron, a detachment of which joined the 2nd Tactical Air Force in Europe, later to be joined by 504 Squadron. The first operational flight was made on April 16 1945, the role being largely ground attack in support of the armies advancing into Germany, owing to the scarcity of any Luftwaffe opposition.

Shortly after the end of the war in Europe, Fighter Command could boast two complete jet fighter Wings: the first, based at Bentwaters comprised Nos 56, 74 and 245 Squadrons; Nos 222, 234 and 263 made up the second, based at Boxted.

Continued improvement in engine design and performance led to the Derwent V which powered the F Mk 4 (in the post-war style of designation) and these engines were installed in more smoothly streamlined nacelles than those of previous marks. In addition, another major appearance change came with the Mk 4 in that the wings were 'clipped' to provide a better rate of roll. Meteor F4s entered RAF service in 1948. By that date the Meteor had made an impression on a wider scale by improving the world's absolute speed record. In the latter part

Meteor F Mk 4 VT170 *on flight test before delivery to the RAF* (MAP).

of 1945 two F3s had been modified to F4 standard and given a highly polished finish to reduce skin friction. These made their attempts on November 7, *EE454* achieving a speed of 606.379 mph at Herne Bay, Kent, with Group Captain H.J. Wilson at the controls. However, the wings on these aircraft were not of the shorter span introduced on the F4s proper and a further attempt was made in September 1946 with F4s at Tangmere, when one of three aircraft involved, *EE549* flown by Group Captain E.M. Donaldson, raised the record to 615.778 mph on the 7th of that month.

The next major variant to appear was the Meteor T Mk 7 which, developed as a private venture, impressed the Air Ministry to the extent that a specification, T.1/47, was written to cover the type's advancement for service use. It was similar to the F4 but had a lengthened forward fuselage to accommodate the second cockpit and the ability to carry three drop tanks giving extra fuel capacity. The aircraft entered RAF service in 1949, also joining the Royal Navy in the 1950s.

The final fighter version to appear from Gloster was the F Mk 8 which differed considerably in appearance from the F4 which it superseded. It featured a lengthened fuselage with increased fuel tankage, a redesigned tail unit and had Derwent 8 engines. All aircraft were fitted with ejector seats. The first production F8 flew in 1949 and the type became operational with the RAF in June 1950, eventually equipping 31 fighter squadrons, including RAuxAF units. F Mk 8s also saw active service with the Royal Australian Air Force during the Korean War.

The last production Meteors to come from the Gloster factory were the FR Mk 9 and PR Mk 10. The former was based on the F8 airframe fitted with a new nose to carry camera equipment for oblique photography, and the latter was a high altitude photo-reconnaissance version with the FR9 nose but with the early long span wing and the F4 tail unit. The FR9 retained the armament and was intended for low-level work.

Gun armament throughout the fighter series was an installation of four 20 mm Hispano cannon in the forward fuselage, in addition to which the F8 had provision for carrying underwing racks for rocket projectiles.

There were many experimental and test-bed uses of the airframe and numerous overseas sales with license production taking place in Belgium and Holland. Other variants of the Meteor included the FR Mk 5 (one F4-based fighter-reconnaissance machine built, lost on its maiden flight); and the U Mk 15 and 16 target-drone conversions from Mks 4 and 8 respectively. (For further information on the Meteor, see Bill Gunston's *Fighters of the Fifties*, also published by Patrick Stephens.)

de Havilland Vampire and Sea Vampire

The 1941 Air Ministry specification E.6/41 called for a single-seat experimental jet-powered aircraft to be capable of carrying an armament of four 20 mm Hispano cannon and achieving a speed of 490 mph. In 1942 detailed design was begun by de Havilland on an aircraft to meet this requirement, the DH 100, at first given the name Spider Crab. The layout chosen was that envisaged earlier by Gloster for the E.28/39 Whittle engine test-bed, in which the fuselage was curtailed in order to install an engine with the power-efficient short jet-pipe and to mount the tail plane high on twin booms to clear the jet efflux. The engine was to be Major F.B. Halford's H.1 (later named Goblin).

Three prototypes were constructed and the first of

The RAF's 'Vintage Pair', still flying in 1982—Meteor T Mk 7 WF791 *and Vampire T Mk 11* XH304 (MoD).

these, *LZ548/G*, was flown for the first time on September 20 1943 from Hatfield with Geoffrey de Havilland, Jr, in the cockpit. The third prototype was the first to have the cannon armament installed. A production order for 300 Vampire F Mk Is (as the type was officially named) was placed with the English Electric Company at Preston, Lancashire, and the first of these flew in April 1945. The type was thus too late to enter war service, the first squadron, No 247, receiving its machines in 1946. A Vampire Wing was formed at Odiham from 247 Squadron together with 54 and 72 Squadrons in September.

On December 3 1945 Lieutenant Commander E.M. Brown made the first carrier landing of a jet aircraft with the second Vampire prototype, *LZ551*, on HMS *Ocean*. A series of trials was carried out over a two-day period and these were so successful that the development of a naval version was ordered, this materialising as the Sea Vampire F Mk 20 in 1948.

The Vampire Mk II was a Rolls-Royce Nene-powered version which did not enter British service but an example shipped out to Australia was used in the development of the Vampire F Mk 30 in that country. The next production version for the RAF was the F Mk 3 which had an increased fuel capacity with provision for drop tanks and a redesigned tail unit. This entered service in 1948. In July of that year, six aircraft of 54 Squadron made the first crossing of the Atlantic by jet aircraft with refuelling stops in Iceland and Greenland.

After the Mk 3 came the FB Mk 5 with the main role being ground attack, provision being made for underwing mounting of bombs or rockets. Aircraft of this mark were used against the Communist terrorists in Malaya in the early 1950s. This was the most widely used version, serving with 39 squadrons of the RAF and RAuxAF. Experience with the type in the Mediterranean and Far East led to the development of a variant specifically for the hot climates of those areas. Designated FB Mk 9, it was fitted with a Godfrey refrigeration unit and air-conditioned cockpit. Other variants of the foregoing series included the Sea Vampire F Mk 21, a strengthened F Mk 20 for undercarriage-less deck-landing trials; the Vampire FB Mks 31 and 32, Australian developments rebuilt from FB Mk 30s; and the Vampire FB Mks 6, 50, 51 and 52 which were export versions. There was also license production of Vampires in Switzerland, Italy and France, the latter country developing the SE535 Mistral (Vampire FB Mk 53) from the FB Mk 5.

On August 28 1949 the prototype of a private venture design, the DH 113, first flew. A development of the Vampire Mark 5, this aircraft was, in essence, a mating of that aircraft's basic structure and engine with the nose and cockpit of the Mosquito night fighter. It was hoped that export orders would be obtained and in October Egypt placed an order for 12. Owing to the political unrest in the eastern Mediterranean an embargo was placed on the sale of arms to Egypt and the aircraft were taken over by the RAF to bolster the country's night fighter force while the transition from Mosquito to Meteor took place. The Vampire NF Mk 10 (as it was designated) entered service with 25 Squadron in July 1951 and equipped only two other squadrons, 23 and 151. The armament remained as for other Vampire marks, ie, four 20 mm Hispano cannon, and the radar was the AI Mk 10 as used in the Meteor NF11. The type was exported to Italy and India with the designation NF Mk 54.

A development of the night fighter was to be the last variant of the Vampire to appear, the T Mk 11. This trainer, which basically involved the omission of the radar and two of the guns and the installation of dual controls, was to be the first jet trainer to enter RAF service. The prototype made its maiden flight on November 15 1950 and production aircraft

Top *Vampire FB5* (Mike Bowyer). **Above** *Vampire NF 54, the export version of the NF 10 supplied to the Italian Air Force* (British Aerospace via Bill Gunston).

Sea Vampire T Mk 22 photographed in 1955 (Mike Bowyer).

entered service in 1952. The Royal Navy also used the type with the designation Sea Vampire T Mk 22. An export model was the T Mk 55 and Australia built the T Mks 33, 34 and 35. (For further information on the Vampire, see Bill Gunston's *Fighters of the Fifties*, also published by Patrick Stephens.)

de Havilland DH 108

Before the Second World War had ended, thoughts had turned to the development of suitable commercial aircraft for use when the conflict was over. The Brabazon Committee had been set up to consider proposals for such aircraft and amongst these was a design from de Havilland for an advanced swept wing tailless jet airliner with the designation DH 106. Owing to the complete lack of practical knowledge of such a layout it was necessary to build a small research aircraft to investigate its performance. The resultant machine, given the designation DH 108 and built to specification E.18/45, used the standard fuselage of the Vampire fighter then in production, fitted with wings of 40° sweepback and a swept fin and rudder. No tailplane was fitted although consideration had been given to this feature in the design. The aircraft was to be powered by the de Havilland Goblin engine.

The first prototype was ready for initial tests at the end of April 1946 and after taxiing trials at Hatfield it was taken by road to RAF Woodbridge to make its first flight. This was achieved on May 15 by Geoffrey de Havilland, Jr, and marked the first flight by a British swept wing jet aeroplane. The second aircraft flew in August and it was the purpose of this machine to explore the high speed performance of the design. It soon proved capable of attaining speeds in excess of the world speed record set on September 7 by a Meteor and it was planned that the DH 108 would be flown over the official course at Tangmere. Tragically this never happened as on the day before the trial was to take place the aircraft was lost over the Thames Estuary during a practice run, the pilot, de Havilland, being killed when the aircraft broke up in the air.

The third DH 108, VW120, photographed in 1948 (Mike Bowyer).

Despite this setback, a third machine was ordered to continue the research work and this embodied refinements including the fitting of an ejector seat. John Cunningham took it on its first flight on July 24 1947. Of major importance in British aviation history was the flight made on September 6 1948, when John Derry, in the third aircraft, achieved a speed in excess of Mach 1.0 in a hair-raising dive from 45,000 ft, during the course of which the machine was at times out of the pilot's control. Derry carried out further flights to investigate speeds in this region in similar circumstances, an effort of dedication to the advancement of flying second-to-none.

The two DH 108s were sent to the RAE Farnborough for further research flights but both were to be lost in fatal accidents, the third machine in February 1950 and the first in May the same year.

Supermarine Attacker

The Attacker stemmed from specification E.10/44 calling for an experimental machine using the newly developed Rolls-Royce Nene jet. Given the Supermarine type number 392, the aircraft built to this specification was based on the wing design of the Spiteful, a piston-engined development of the Spitfire series, with a new fuselage to accommodate the Nene engine of near circular section and with lateral air intakes alongside the cockpit. Owing to the simplicity of layout, the first aircraft was completed by July 1946 and made its first flight on the 27th of that month at Boscombe Down in the hands of chief test pilot Jeffrey Quill.

The RAF, satisfied with the Meteor and Vampire, had not shown any interest in development of the Type 392 for service use, but the Royal Navy was interested and 'navalisation' of the design to specification E.1/45 resulted in the appearance of the Type 398. Lieutenant Commander Mike Lithgow took the new machine on its maiden flight on June 17 1947 and the first deck landings were made on HMS *Illustrious*, the trials carrier, on October 5.

It was not until November 1949 that an order was placed for the Attacker—the name had been applied early in 1947—and the first aircraft from the production batch flew in April 1950. The initial service versions were the F Mk 1 and FB Mk 1, the armament of which was four wing-mounted 20 mm cannon in both marks plus underwing positions for eight rocket projectiles or two 1,000 lb bombs in the FB Mk 1. The first squadron to receive the Attacker was No 800, based at Ford, Sussex, in August 1951. Only three front-line squadrons were destined to fly the type and in 1954 replacement by the Hawker

Supermarine Attacker FB2 WZ273, *the first aircraft in the final production batch of 30; 85 were built in all* (Mike Bowyer).

Sea Hawk commenced. The Attackers thus relinquished from first-line service equipped five squadrons of the RNVR until 1957.

The only other mark was the FB Mk 2 in which the Nene 3 engine was replaced by the Nene 102 and a new cockpit hood was fitted. All service aircraft were capable of receiving RATOG (rocket-assisted take-off gear), and were fitted with Martin-Baker ejector seats. The Royal Pakistan Air Force ordered land-based versions of the F Mk 1 and these were delivered in 1951.

Saunders-Roe SR.A/1

As the tide of the Pacific war turned and the Allied forces began their island-hopping campaign across that ocean, a requirement appeared for a flying boat fighter of high performance, which could take on the leading Japanese combat aircraft such as the Mitsubishi A6M Zero. A water-borne aircraft would be able to advance as the islands were retaken without having to wait for advanced airstrips to be cleared. The major drawback of the flying boat, however, was that owing to the necessity of a planing bottom an ideal aerodynamic form was impossible to achieve. This, connected with a piston engine, would mean that a performance adequate to combat the Japanese fighters would be hard to come by.

The Saunders-Roe company saw a solution to the problem in utilising the jet engine to enable the optimum planing bottom fuselage to be designed without the constraint of a piston engine and the high surface clearance needed for its propeller. In 1944 work was begun, to specification E.6/44, on a flying boat fighter to be fitted with two Metropolitan-Vickers F.2/4 Beryl axial-flow turbojets. The armament was to be four 20 mm Hispano cannon nose-mounted above the air intake.

Although the end of the war removed the immediate need for the SR.A/1, it was hoped that export orders would be received and construction of three prototypes was ordered. The aircraft were the product of two islands, component manufacture taking place at Beaumaris, Anglesey, and assembly at the Cowes works on the Isle of Wight. The first machine made its maiden flight on July 16 1947 at Cowes with Geoffrey Tyson as pilot, who two months later demonstrated the aeroplane at the SBAC show at Farnborough. In 1948 the two other aircraft joined the flight test programme but misfortune befell the first of these when it crashed

The SR.A/1 taxying outside Cowes (British Hovercraft Corporation Ltd via Mike Bowyer).

into the sea on a flight from the MAEE Felixstowe. The third prototype was also lost after hitting a baulk of timber while alighting off Cowes. The remaining prototype continued to fly sporadically until 1951, in which year it was displayed on the Thames as part of the Festival of Britain.

No orders for this unique marine aircraft had been received but fortunately for posterity the survivor became a museum exhibit, at present part of the Imperial War Museum collection at Duxford.

Hawker P.1040, P.1072, P.1052 and P.1081

The first Hawker jet was constructed to meet specification N.7/46 for a naval interceptor, with power from a Rolls-Royce Nene engine. The first flight was made at Boscombe Down on September 2 1947 by Squadron Leader T.S. Wade. No armament or 'naval' features (ie, folding wings/arrestor hook) were fitted. The third aircraft was the true prototype of the Sea Hawk. The first P.1040 was modified, to become the P.1072, by the addition of an Armstrong Siddeley Snarler rocket engine in the tail. The first flight in this form was made on November 20 1950 but trials were abandoned owing to the Air Ministry's preference for afterburner development.

The P.1052 was a swept main-plane derivative of the P.1040 to specification E.38/46, powered by the Nene 2 engine. The first flight was carried out by Squadron Leader Wade from Boscombe Down on November 19 1948. The aircraft was used in combat manoeuvre trials with a Meteor Mk 4. In May 1952 the same machine made deck landings on HMS *Eagle*. The second aircraft was converted into the P.1081.

Originally to have been powered by a Rolls-Royce Tay, the P.1081 emerged as a Nene-engined aeroplane with a new tail jet-pipe outlet and a swept fin and tail plane. Squadron Leader Wade again was responsible for the first flight, this taking place on June 19 1950. Australian interest led to the P.1081 but when that interest waned the aircraft was handed over to the RAE Farnborough for trials. On April 3 1951 Wade was killed when the machine crashed, the cause of the accident never being determined.

Right, top to bottom *Hawker P.1040 photographed in 1948. P.1072, which was used for Snarler rocket trials. Navalised P.1052* VX272. *P.1081* VX279 *photographed in July 1950* (Mike Bowyer).

Armstrong Whitworth AW 52

In 1944 work began at the Armstrong Whitworth factory at Baginton, Coventry, on a tailless glider, designated AW 52G, the purpose of which was to obtain information for a proposed six-jet airliner of similar overall configuration. This first flew on March 2 1945. To gain high-speed experience a powered aeroplane, again tailless, was designed to Air Ministry specification E.9/44 and this was to include a laminar-flow wing section.

Two prototypes were built, one fitted with a pair of Rolls-Royce Nene engines and the other with two Derwents. The first, Nene-powered, prototype was completed in 1947 and taken to the A&AEE at Boscombe Down for flight trials which began on November 13. The second aircraft first flew in September 1948. During a flight on May 30 1949 the pilot ejected from the first prototype when an uncontrollable pitching oscillation occurred due to wing flutter. The machine subsequently crash-landed in open country and, although not seriously damaged, was not repaired. The second aircraft continued to fly and in 1950 went to Farnborough to take part in further tests.

The necessity to maintain a spotless finish of the wing surfaces to achieve the optimum laminar-flow performance was found to be impracticable and would have been impossible to achieve in everyday use of a production aircraft. The original scheme for the six-engined airliner was abandoned and the remaining AW 52 was scrapped in 1954.

Official Armstrong Whitworth photo of the strange looking AW 52 (via Mike Bowyer).

Gloster E.1/44

With construction of the twin-engined Meteor under way, consideration of single-engined designs was not forgotten. An engine of sufficient power to enable a respectable performance to be achieved in a single installation would become available as the importance of the jet engine for the future ensured an increasing pace of technical development. In November 1943 work began on two prototypes to specification E.5/42 (modified) to take the Rolls-Royce Nene engine, a larger development of the Derwent promising greater thrust. However, work on these airframes was stopped when a new specification, E.1/44, was issued and a third machine was begun. This reached completion in July 1947 but had the misfortune to be severely damaged in an accident while being taken by road to Boscombe Down for flight trials. Eventually a fourth airframe was completed and flight achieved for the first time on March 9 1948 at Boscombe Down. It was discovered that handling of the machine was not satisfactory and a new tail was designed. This was fitted to the fifth airframe and flight trials in 1949 proved it a success. This was the same tail unit as that used on the Meteor F8 then starting production.

A planned installation of four 20 mm Hispano cannon under the forward fuselage is unlikely to have been fitted to the aircraft as it was decided that further development was not warranted in view of the versatility of the Meteor. A sixth airframe, still under construction, was abandoned and the two extant aeroplanes spent the remainder of their

The first Gloster E.1/44 GA 2 with original tail and anti-spin parachute housing above the jet-pipe (MAP).

careers as trials aircraft at the RAE Farnborough. The six aircraft received designations in the Society of British Aircraft Constructors' series instigated in 1946. GA 1 was applied to the first two (uncompleted) airframes, GA 2 to the next three, GA 3 to the sixth (unbuilt) and GA 4 was allotted to the proposed production aircraft.

Vickers Nene Viking

The Type 618 was a one-off conversion from the standard Vickers Viking airliner in which two Rolls-Royce Nene turbojets were installed in underwing nacelles in place of the usual power-plant of the two Bristol Hercules piston engines. A new undercarriage was designed and airframe construction was modified to meet the demands of a near-doubling in air speed. The first flight was made from Wisley on April 6 1948 and, on July 25 the same year, a pioneer passenger-carrying flight was made from Heathrow Airport to Villacoublay (Paris) in 34 min 7 sec, considerably faster than the contemporary scheduled flight times. Flight trials continued for several years, obtaining data for the forthcoming jet airliners until, in 1954, the aircraft was rebuilt as a standard Viking freight carrier.

The Nene Viking, G-AJPH (via Richard Riding).

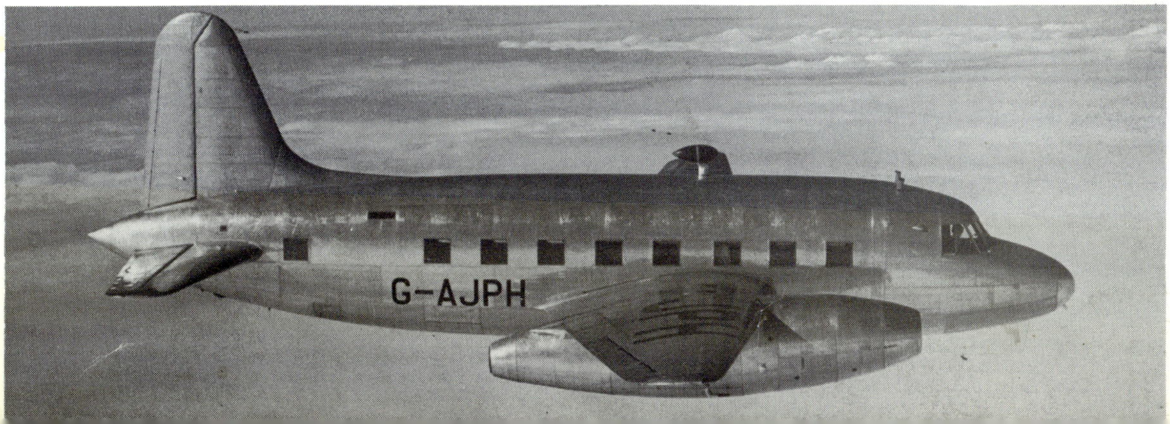

Supermarine 510, 535 and 541

The Type 510 was an experimental aircraft to specification E.41/46 using an Attacker fuselage fitted with a swept-back wing and swept tail surfaces of 40° at quarter chord. It retained the tailwheel undercarriage. *VV106* first flew on December 29 1948 and, on November 8 1950, made the first deck landing by a swept wing aircraft on HMS *Illustrious*. The pilot on that occasion was Lieutenant D.G. Parker, RN.

The Type 535 was originally the Type 517 (a Type 510 with afterburning Nene), redesignated after fitting a tricycle undercarriage and longer nose. It was first flown on August 23 1950.

The Type 541 was a refinement of the Type 535 design, fitted with a Rolls-Royce Avon engine which made its first flight on August 5 1951. This was the immediate forerunner of the Supermarine Swift.

Above right and right *Two views of Supermarine 510* VV106 *photographed in July 1950* (Mike Bowyer).

Below *Supermarine 535* VV119 (Rolls-Royce via Mike Bowyer).

English Electric Canberra

The origins of the long-lived Canberra lie in schemes for a jet fighter-bomber evolved in 1944 by W.E.W. Petter of Westland Aircraft Ltd. The English Electric Company of Preston, Lancashire, had, throughout the war, built numerous aircraft, particularly the Hampden and Halifax bombers designed by Handley Page, but had not possessed their own design department. However, as the war's end approached, the company set up a design staff and Petter left Westland to become its chief engineer.

Their first major project was for a light jet bomber evolved from the ideas Petter had been working on at Westland. This was a design made to Ministry of Supply specification B.3/45 to be powered by two Rolls-Royce axial-flow jets then under development. A contract was placed in January 1946 for four prototypes. Designated EE A.1, the first of them was ready for initial trials in May 1949. Taxiing began at Warton (an ex-USAAF base) on the 7th of that month, the first hop was made on the 9th and on the 13th (a Friday) the first flight was successfully completed by Wing Commander R.P. Beamont. Thereafter flight trials proceeded smoothly with the three other prototypes all flying before the end of the year. The second machine was fitted with Nene engines as a security against failure of the new Avons in the other three.

It had been intended that the type, by now named Canberra, would have a crew of two, a radar bombing system in the nose and capacity for a 6,000-lb bomb load. Difficulties with the radar equipment led to a revision in the production version to include a third crew member for visual bomb aiming. The specification B.5/47 was issued to cover this and the designation B Mk 2 given, the four prototypes being representative of the B Mk 1. The first production aircraft made its maiden flight on October 8 1950 and deliveries began, to No 101 Squadron, in May 1951.

The following is a summary of the many succeeding versions of the Canberra.

PR Mk 3 Photographic reconnaissance version to specification P.R.31/46 to replace Mosquitoes. Incorporated a lengthened fuselage to accommodate the cameras. A Mk 3 won the London to New

VN813, *the second Canberra B 1 prototype, was used by Rolls-Royce 1950-51 for Nene engine development* (Rolls-Royce via Bryan Philpott).

Zealand Air Race in October 1953 covering the 12,270 miles in 23 hr 50 min 42 sec.

T Mk 4 First trainer variant to specification T.2/49.

B Mk 5 Proposed target marker to specification B.22/48. One prototype only.

B Mk 6 Improvements over Mk 2 included higher powered engines and greater fuel load for increased range. Mk 6s used in action against terrorists in Malaya from 1955 and during the 1956 Suez crisis.

B (I)Mk 6 Interdictor version of Mk 6 with belly pack of four 20 mm guns and underwing bomb/rocket pylons. Used by the RAF in Germany, equipping only 213 Squadron.

PR Mk 7 Photo-recce version of Mk 6 similar to PR3.

B (I)Mk 8 Represented a major advance over previous 'B' marks with a redesigned forward fuselage, in which the pilot occupied a fighter-type cockpit offset to port and the navigator/bomb aimer's position was in the nose, there being only the two crew members. Armament included a ventral gun pack of four 20 mm cannon in the rear section of the bomb bay and underwing bomb points capable of taking two 1,000-lb bombs or AS 30 missiles in addition to three 1,000-lb bombs in the bay. Provision for fitting a camera pack was made in the forward section of the bomb bay.

PR Mk 9 High altitude photo-recce version with new engines and increased wingspan. Production undertaken by Short. Offset canopy as in B (I)Mk 8.

U Mk 10 Pilotless aircraft for use in missile trials by the Royal Navy, modified from B Mk 2s by Short.

T Mk 11 Trainer variant with extended nose containing AI radar and accommodation for a crew of four, including two students. Converted from Mk 2.

Standard production Canberra B 6, WJ764, *of 101 Squadron* (BAC via Bryan Philpott).

B (I)Mk 12 Export model for New Zealand.

T Mk 13 Export model for New Zealand.

D Mk 14 Drone aircraft; improved version of U Mk 10 converted from Mk 2.

B Mks 15 and 16 Modified from ex-Bomber Command B Mk 6s for the Near East and Far East Air Forces, with provision for cameras in the nose and underwing rocket pods. Additional radio equipment fitted.

E Mk 15 Radar and radio calibration version converted from B Mk 15 after withdrawal of RAF overseas units.

T Mk 17 Conversion (again from Mk 2) for electronic counter-measures training.

TT Mk 18 Conversion for target towing duties for RAF and RN.

T Mk 19 T Mk 11s with AI radar removed.

T Mk 22 Conversion for RN use in radar operator training. Unlike all previous conversions which used retired Mk 2s, this version was from PR Mk 7s.

In addition to the British-built marks listed above, license production took place in Australia. Two B Mk 2s were used as pattern aircraft by the Government Aircraft Factory, whose production version was designated B Mk 20. Aircraft converted to dual control became T Mk 21s.

America was impressed by demonstrations of the prototype Canberras and following trials in America against indigenous designs, an agreement on license production was made in 1951, resulting in a parallel development as the Martin B-57.

Several export orders were obtained for the Canberra, starting in the early 1950s with Venezuela; other countries being France, Ecuador, Peru, India, Rhodesia, South Africa, Germany, Ethiopia and Argentina. Mostly these were aircraft of marks listed above, eg B (I)8s, PR3s and T11s,

Canberra PR 7 WH799 *of 13 Squadron which was shot down over Syria in 1956* (English Electric via Bryan Philpott).

Canberra T 4 in natural metal finish with yellow trainer bands (MoD).

but some machines have had new designations including B (I)Mk 58 for Venezuela and India, PR Mk 57 for India, B Mk 66 for Peru, B (I)Mk 68 for Peru, T Mk 64 for Argentina, B Mk 62 for Argentina and B Mk 52 for Ethiopia.

de Havilland Comet

To begin with the Comet (DH 106) had been envisaged as a tailless aircraft, as mentioned in the section on the DH 108 experimental aircraft, but even before that aeroplane had flown it was decided to use a conventional layout for this pioneer commercial jet. The definitive design to Ministry of Supply specification 22/46 was begun in September 1946 and work on the first two prototypes commenced the following year. On July 27 1949 the maiden flight was made by John Cunningham from Hatfield. Production orders had already been received—nine aircraft for BOAC—and the first of these flew on January 9 1951. The Comet 1 was powered by four de Havilland Ghost engines, a development of the Goblin, which had been extensively test flown in Lancastrian test-bed aircraft and in a modified Vampire for high altitude trials, in the course of which an altitude record of 59,446 ft was achieved. This boded well for the Comet's service career, and deliveries to BOAC began in April 1952 with the first fare-paying passenger-carrying flight quickly following on May 2, this being from London (Heathrow) to Johannesburg in six stages. Other orders for the Comet had been received from Canada and France and these export machines were 1As with seating capacity increased from 36 to 44 and greater fuel load giving more range.

The next version, the Comet 2, was powered by Rolls-Royce Avon axial-flow engines giving more power than the centrifugal Ghosts, and these had a three-foot fuselage extension allowing a further increase in seating. Before these aircraft could enter service a series of accidents involving Comet 1s had caused a major setback to the airliner's career. As a result the production of the '2 was stopped after 18 had been completed and of these 13 were modified for use by the RAF. Two, designated T Mk 2, were used for training and eight as C Mk 2s served as military transports, the two trainers being later altered to C Mk 2 standard. The other three C Mk

Above *The Comet 1 prototype* G-ALVG *photographed at Farnborough in 1950* (Mike Bowyer).

Right and below right *BOAC Comet* G-ANLO, *and* SU-ALD *(of United Arab Airlines)* (Mike Bowyer).

2s were used on special duties work. Deliveries to the RAF began in mid-1956 and the type remained with Transport Command until 1967.

Only one example of the Comet 3, with up-rated Avons, lengthened fuselage and additional pod fuel tanks on the leading edges of the wings, was constructed. During an extensive period of development flying in 1954-55 this aircraft became the first jet airliner to circumnavigate the world.

Exhaustive research at the RAE Farnborough showed that explosive decompression had been responsible for the loss of two of the Comet 1s, metal fatigue having started at the corner of one of the square cabin windows. As a result all subsequent aircraft were fitted with oval windows. Although the existing aircraft could have been modified, this course was not followed and a new model, the Comet 4, appeared.

The Comet 4 was developed from the '3 and the first aircraft of an order for 19 from BOAC first flew on April 27 1958. The first transatlantic fare-paying passenger-carrying flights were made on October 4 that year when two aircraft flew simultaneously on east and west bound routes. Numerous foreign airlines placed orders for the Comet 4 series and several remain in service. The maximum number of passengers that could be carried increased from 81 in the Comet 4 to 101 in the '4C. The RAF took delivery, in 1962, of five Comet C Mk 4s to supplement its Mk 2s and these remained in service until June 1975 when the transport fleet was cut back.

de Havilland Venom and Sea Venom

The plan to re-engine the Vampire with the de Havilland Ghost led to a redesign of the airframe under the company designation DH 112, including a completely new wing of thinner section and different plan to exploit the extra power provided. Wingtip tanks, which could be jettisoned in flight, were provided in place of the then standard pylon-mounted underwing drop tanks.

At first designated Vampire FB Mk 8, the type was soon renamed Venom and ordered against specification F.15/49 as a Vampire replacement. The prototype had been flown for the first time on September 2 1949 from Hatfield and soon proved its superiority in performance over the Vampire and Meteor in trials carried out at the A&AEE Boscombe Down. Production machines carried an armament of four 20 mm Hispano cannon under the nose and could take underwing loads of 2,000 lb of bombs or eight 60-lb rocket projectiles. The first Venon FB Mk 1s to enter service joined the 2nd

Venom 1 prototype VV612 (de Havilland via Mike Bowyer).

Tactical Air Force in Germany in 1952 and two years later began equipping squadrons of the Middle East Air Force. The FB Mk 4, with power-operated ailerons and redesigned fins and tailplane, appeared in 1955 and this mark was fitted with an ejector seat. Aircraft of both marks took part in the Malayan jungle war against the Communist terrorists, and the Suez crisis of 1956 saw Mk 4s involved flying from bases in Cyprus.

As with the Vampire night fighter, a private venture machine led to the appearance of the Venom NF Mk 2 which incorporated a new fuselage similar to that of the Vampire NF Mk 10 with the wings and tail structure of the Venom Mark 1. The prototype first flew on August 22 1950 and the first production machine on March 4 1952, but time taken to solve handling problems delayed entry into service until May 1953. The NF Mk 3 incorporated airframe changes and was fitted with the American-designed AI Mk 21 radar in place of the Mk 10 of the earlier aircraft. These entered service in mid-1955.

Venom variants for overseas sale were the FB Mk 50 for Iraq, NF Mk 51 for Sweden and FB Mk 4 for Venezuela. License production involved Switzerland (whose Venoms soldiered on into the 1980s) and France where the type was built as the Sud-Est Aquilon. This latter type evolved from the Sea Venom with the export designation NF Mk 52, although none of that mark was built as such.

DH 112 Sea Venom FAW 20 prototype WK379 *did not have folding wings* (de Havilland via Mike Bowyer).

The Royal Navy had evaluated the NF Mk 2 prototype and following on from that had ordered a 'navalised' prototype, for further trials, to specification N.107. This was at first known as the Sea Venom NF Mk 20 and carrier trials started on HMS *Illustrious* in July 1951. Production aircraft, designated FAW Mk 20, first appeared in 1953 and started joining the squadrons in March 1954. The FAW Mk 21 embodied modifications similar to those in the Venom NF Mk 3, added to which was a long-stroke undercarriage and provision for fitting RATOG. Inclusion of an uprated engine led to the designation FAW Mk 22. Carrier-borne Sea Venoms assisted the Cyprus-based RAF Venoms during the Suez actions.

For the Royal Australian Navy a version of the Mk 21 was built in which minor equipment variations were made to suit that service and this was designated FAW Mk 53.

Avro 707

The specification B.35/46 which called for an advanced bomber using the latest technical knowledge (particularly that of wartime Germany) resulted in two designs which eventually entered service as the Vulcan and Victor. Owing to the nature of these designs the Ministry of Supply ordered a series of small research machines to be developed in parallel with the bomber prototypes to prove their layouts. In the case of the Avro entry, the Vulcan, these took shape as the Avro 707 delta wing series.

The first aircraft, built to specification E.15/48, was started at the end of 1948 and used standard main undercarriage components from the Athena with Gloster Meteor nose gear and cockpit canopy to speed up construction and get it into the air as quickly as possible. The Rolls-Royce Derwent engine was used, the air intake for which was positioned midway along the top of the fuselage to leave the wing a clean delta. (In the Vulcan the four engines were to be buried in the wing with leading edge air intakes.) The 707 was taken to Boscombe Down for testing and the maiden flight was made on September 4 1949. It appeared at the SBAC show at Farnborough but before the month was out it was destroyed in a crash near Blackbushe, killing the test pilot, Eric Esler.

In order that trials could continue two further aeroplanes were ordered, the first of these designated 707B with modifications to the original design, and the second to the new specification E.10/49, for high speed research featuring further refinements including wing leading edge air intakes. This machine was to be the 707A. The 707B first flew on September 6 1950 and the 707A on June 14 1951. This was too late to allow any data to be applied to the Vulcan prototype which was then in an advanced state of construction but flying went ahead to prove the 707 itself and to continue basic research on delta wing performance.

Two further aircraft were built, a second 707A and a two-seater, the 707C. The latter, four of which had been ordered, was planned as a side-by-side dual control trainer for future delta pilots. The 707C first flew on July 1 1953. This machine, in the early 1960s, made a series of trials of the 'fly-by-wire' system of control in which the control surfaces are operated by electrical signals from the 'joy-stick' rather than by a mechanical linkage.

The one and only Avro 707B, VX790 (Avro via Mike Bowyer).

Hawker Sea Hawk

Following the first two P.1040 prototypes (see page 17) came a third machine, *VP422*, which was equipped with most of the production version's refinements including a lengthened arrestor hook and provision for underwing drop tanks and RATOG. This machine first flew on October 17 1949 and, with the second P.1040, carried out initial service trials in 1950.

The first production batch of Sea Hawks, F Mk 1s, was begun in 1951 at the Kingston factory and these had a 2 ft 6 in increase in wingspan to allow for additional equipment weight, including a Martin-Baker ejector seat. It soon became clear that Hawker's facilities were going to be unable to cope with production of Sea Hawks and Hunters, both of which were accorded 'super priority' status, owing to concern that a Russian attack on the West, at the time of the Korean War, was planned for the mid-1950s. Therefore the future production of the Sea Hawk was handed over to the Armstrong Whitworth factory at Baginton, that company subsequently building all further examples of the type.

Carrier trials with the second P.1040 had been made on HMS *Illustrious* in April 1949, these leading to the discovery of the need for the lengthened arrestor hook, and the first production aircraft trials were made on HMS *Eagle* in 1952 using two of the 35 built by Hawker before the production hand-over. The first squadron to be equipped was No 806 based at Brawdy in March 1953, the first embarkation being made in February the following year aboard *Eagle*.

Subsequent variants of the Sea Hawk were as follows: the F Mk 2 which introduced powered aileron control to improve lateral stability; the FB Mk 3 with provision for carrying, underwing, two 1,000-lb bombs in place of long-range drop tanks; and the FGA Mk 4 ground attack version with

The graceful lines of the Sea Hawk F Mk 1 (via Mike Bowyer).

provision for carrying two 500-lb bombs plus racks under the outer wings for rocket projectiles. Substitution of the uprated Nene 103 engine in the last two marks mentioned produced the FB Mk 5 and FGA Mk 6; an additional batch of Mk 6s was built as such and these were the last Sea Hawks for the Royal Navy, delivery being completed by January 1956.

Gun armament in all marks was a battery of four 20 mm Hispano cannon mounted in the underside of the forward fuselage and 21 squadrons were equipped with one or more of the six marks, including three Volunteer Reserve units and five training squadrons. Of the front-line squadrons, six saw active service during the Suez crisis in 1956.

Sea Hawks also entered service with three foreign countries, the Netherlands, Germany and India, as the FGA Mk 50 which was identical to the FGA Mk 6 but with new radio equipment; the Mk 100/101 West German Navy versions, for fighter-bomber/ground attack and radar reconnaissance respectively; and the FGA Mk 6—ex-RN aircraft supplied to the Indian Navy plus a batch in 1961, the last Sea Hawks to be built. (For further information on the Sea Hawk, see Bill Gunston's *Fighters of the Fifties*, also published by Patrick Stephens.)

Late-production Sea Hawk FGA 6 of 804 Squadron (via Bill Gunston).

Vickers Tay Viscount

Having modified a Viking airframe to take a pair of Rolls-Royce Nene engines, a similar conversion was made to an early Viscount resulting in the Type 663. The engines this time were Rolls-Royce Tays, a development of the Nene. The first flight was made from Wisley on March 15 1950. The aircraft's main work was in the test flying of new control systems, initially that designed by Boulton Paul for

Originally Viscount G-AHRG, VX217 *was the Type 663 testbed for the Tay engine* (Mike Bowyer).

the Vickers Type 660 Valiant bomber. Subsequently the first 'fly-by-wire' trials were carried out by the same machine.

Armstrong Whitworth Meteor

As the 1950s approached, the RAF night fighter force was still equipped with the de Havilland Mosquito which, fine aircraft though it was, was obsolete and out-performed by the B-29 bombers already in *Russian* service. Several companies had submitted designs for two-seat jet aircraft with night and all-weather capability but these were deemed, by the Air Ministry, to be unsuitable. As an interim measure, therefore, it was decided to develop the Gloster Meteor to fill the gap until such time as new aircraft could be designed and built for the purpose. At that time (1949) the Gloster company was heavily involved with production of the Meteor F8 and the design work for the GA 5 which emerged as the Javelin. Armstrong

AW Meteor NF 12 with American radar, enlarged fin and Derwent 9 engines (British Aerospace via Bill Gunston).

Whitworth were already building Meteors and the responsibility for development of the night fighter was handed over to them.

A T7 airframe was modified to test the aerodynamic form of the new design and first flew in October, with the first prototype NF Mk 11 (as the type was designated) flying on May 31 1950. The Meteor night fighter was to meet Air Ministry specification F.24/48 and differed from its predecessors in several ways. The wings used were of the greater span of the Mk I/III; the fuselage had an extended nose to house AI Mk 10 radar, resulting in the guns being moved to the wings outboard of the engine nacelles; the cockpit was modified to include the radar displays for the operator and pressurised to allow for high altitude interceptions; and the tail used was that of the F8. No 29 Squadron was first to become equipped with the type, in August 1951 at Tangmere. A development of the NF 11 was the NF Mk 13 which had minor adaptations to suit the aircraft for service in hot climates; it entered service with two squadrons of the Middle East Air Force.

The NF Mk 12 was a further development with a nose of increased length to accommodate the American APS 21 radar and slightly increased fin area. The first flight was made on April 21 1953 and aircraft entered service later that year. The final mark of the night fighter, and the last new Meteor model to appear, was the NF Mk 14, first flight being made on October 23 1953 with deliveries to the RAF beginning in March 1954. This version had a clear-view cockpit canopy as opposed to the heavily framed hood of the preceeding aircraft and a nose of even greater length.

Modified from ex-RAF NF 11s were the Meteor TT Mk 20s for the Royal Navy, which were equipped to carry towed targets for gunnery and guided-missile practice. Several night fighter Meteors were used in engine and weapon trials and sales of the aircraft made overseas. One of these, in French hands, had the honour of ushering in the age of the supersonic transport, by acting as chase 'plane during the first flight of Concorde from Toulouse on March 2 1969.

Avro Ashton

The Ashton evolved from the Avro Tudor series via the Tudor 8. This was, originally, the second prototype Tudor Mk 1 which received a major rebuild in 1948 and, with a powerplant of four Rolls-Royce Nene engines in two underwing twin nacelles, became the first British four-jet aircraft to fly, on September 6 1948. It was used for high-altitude research at Boscombe Down. The next step towards the Ashton was the Tudor Mk 9, a proposed jet transport with tricycle undercarriage, which in the

The sole Avro Ashton Mk 1, WB490 (Mike Bowyer).

event was never built.

The Ashton design was based on the Tudor 9 and six aircraft were ordered by the Ministry of Supply for various high-altitude research projects. The first of these Avro 706s, the Ashton Mk 1, made its maiden flight on September 1 1950 and was used to gather data on the high altitude performance of turbojets, flying from Boscombe Down. Five other aircraft were built: one Mk 2, three Mk 3s and one Mk 4. The Mk 2 was used for air-conditioning tests, based at RAE Farnborough, and later carried out engine trials with various different turbojets mounted in an under-fuselage pod. The first Mk 3 was used in research on radar bombing systems; the second as a test bed for Bristol Engines, flying with two additional engines in underwing pods; and the third for bombing trials at RAE Orfordness and later as engine testbed in intake icing trials. The final aircraft, the Mk 4, was used by RAE Farnborough as another bombing trials machine with a ventral pannier to carry a bomb aimer and underwing bomb containers.

The six Ashtons worked through the mid-1950s on their varied tasks, the final survivor being the second Mk 3 which lasted until June 1963.

Boulton Paul P.111 and P.120

These two products of the Boulton Paul company were designed purely as delta wing research aircraft, unlike the Avro 707s which were intended to gain information for the Vulcan programme. The P.111 was built to Ministry of Supply specification E.27/46 and, fitted with a Rolls-Royce Nene engine, was to explore the transonic capabilities of the delta wing. Its first flight was made at Boscombe Down on October 6 1950. After a landing mishap early in 1952 the aircraft underwent internal modification with the inclusion of an advanced flying control system, an area in which Boulton Paul were pioneers. In addition four fuselage-mounted airbrakes and a nose probe were added. The P.111A, as it was redesignated, first flew on July 2 1953 and, after being handed over to the RAE, continued its research career until 1958. Thereafter the machine was used for ground training of apprentices at the College of Aeronautics, Cranfield.

On August 6 1952 a revised version of the design, to specification E.27/49, called P.120, had made its maiden flight, also at Boscombe Down. The major difference between the two aircraft was in their tail structure, the P.120 having a swept fin and rudder and high mounted horizontal surfaces of the all-moving slab type. These allowed greater control of longitudinal stability but, only three weeks after the initial flight, control flutter forced the pilot to eject and the aircraft was destroyed.

Top *The P.111* VT935 *photographed in 1951* (Mike Bowyer) *and* **above** *the P.120* (Boulton Paul via Mike Bowyer).

Fairey FD 1

Work, by Fairey, on a series of delta-winged models was initiated in 1946 to obtain data for a proposed manned interceptor to be launched from a sharply inclined or vertical ramp. These rocket-propelled models were used in an extended series of trials carried out on the Aberporth ranges on the west coast of Wales and at Woomera in Australia. Their results varied from total failure to qualified success.

In 1947 the Ministry of Supply issued a specification, E.10/47, for the development of a manned delta wing aircraft to obtain performance

VX350

Facing page *Two views of the Fairey FD 1,* VX350 (Mike Bowyer).

data at high subsonic speeds. Three machines were ordered but only the first of these was destined to be built, the work being done at Fairey's Heaton Chapel factory, Manchester.

The FD1, or Delta One, was intended to take off vertically, as in the case of the models, with power from a Rolls-Royce Derwent turbojet plus four booster rockets. This idea was abandoned in favour of conventional use of a runway, for which purpose a tricycle undercarriage was used, retracting into the fuselage. The rockets were omitted. The aircraft was completed by May 1950 and taxiing trials were made at Ringway Airport; however, it was not until the following year that the 'Delta' was taken, by road, to Boscombe Down to carry out flying trials, the first flight being achieved on March 12 1951 by Group Captain R. Gordon Slade, company chief test pilot.

The aircraft furthered the knowledge of the handling qualities of the delta wing but never approached the planned top speed owing to the horizontal tail surfaces on the top of the fin never being removed, as was intended for high speed flight. During a flight by a service pilot, a landing mishap caused damage to the undercarriage which was not repaired and no further flying was done.

Vickers Valiant

The first operational strategic jet bomber to join the RAF evolved from the requirements laid down by the Air Staff in 1946 for advanced aircraft capable of carrying nuclear weapons over long ranges and delivering them accurately from high altitude using electronic aiming equipment. The three main submissions were from Avro, Handley Page and Vickers and this trio of aircraft were generically termed the 'V-Bombers', although no two sources agree on the name's origin. As it happened, the least radical machine of the three and the first to fly was the last to be ordered as the design was at first considered to be short on range. It was realised, though, that the development time of the Vickers design was likely to be shorter than that required for the Avro and Handley Page submissions and consequently it would be able to enter service so much earlier.

The Ministry of Supply specification B.9/48 was allocated to cover the Vickers Type 660 and two prototypes were ordered on February 2 1949. The first of these was to be powered by four Rolls-Royce Avons and the second, allocated the manufacturer's type number 667, four Armstrong Siddeley Sapphires. The first prototype was erected at Wisley in 1951 from sub-assemblies made at the Foxwarren experimental shop and the maiden flight made on May 18 1951 by J. Summers and G.R. Bryce. The name Valiant was applied in June. Flight testing was brought to a halt the following January following the loss of the first prototype. The aircraft was abandoned in flight, owing to a fire starting in the port wing during engine stop and start trials. However, three months later the second prototype flew for the first time on April 11, powered by Avons instead of the planned Sapphires.

Also in April, the first production order was placed for 25 aircraft; the first Valiant B Mk 1 flying on December 22 1953. To prepare RAF crews to take the Valiant into squadron service an Operational Conversion Unit (No 232) was formed at Gaydon, Warwickshire. No 138 Squadron was formed alongside to become the first operational squadron in February 1955. Thereafter the Valiant force steadily expanded, ten squadrons eventually seeing service on the type.

There were three further variants after the standard B Mk 1, these being the B (PR)Mk 1, B (K)Mk 1 and B PR (K)Mk 1, where 'PR' indicates a strategic reconnaissance capability and 'K' a flight-refuelling tanker role.

Highlights of the Valiant's career include, among many successful exercises and competitions (notably in the USA), the dropping of Britain's first 'A' Bomb over Maralinga, South Australia, on October 11 1956 and the first 'H' Bomb in 'Operation Grapple', while based at Christmas Island in the Pacific, on May 15 1957. During the Suez campaign of October and November 1956, Valiants of four squadrons flying from Luqa, Malta, carried out pinpoint attacks using conventional bombs on varied targets, the first being the airfield at Almaza, Cairo, on October 31.

One other variant of the Valiant appeared, designated B Mk 2, this being specifically for low-level pathfinder missions with structural strengthening to allow for the rougher conditions at these altitudes. A lengthened forward fuselage and wing pods on the trailing edge containing the undercarriage when retracted were features which distinguished the Mk 2. Only one prototype was built and flown, first flight being made on September 4 1953 by G.R. Bryce and Brian Trubshaw, from Wisley. One Mk 1 Valiant had bomb bay modifications to permit trials with the Avro Blue Steel stand-off bomb, although this weapon was never issued to Valiant squadrons.

The realisation of the vulnerability of the big

WZ365

WZ378

bomber to detection and missile attack when at high altitude led to the change, in 1963, of the V-Bombers' role to that of low-level strike. In July 1964 metal fatigue was discovered in the wing spars of the Valiants with the result that the entire fleet was grounded, the decision being taken in January 1965 to scrap.

Left *Vickers Valiant B 1,* WZ365, *at its Farnborough debut in 1951* (Mike Bowyer). **Below left** *Valiant B Mk 1* WZ378 *on a visit to Malta in 1962* (MAP).

Handley Page HP 88

As the Avro 707 series related to the delta Vulcan design, so the HP 88 was to explore the handling qualities of the crescent wing and tail planned for the Handley Page Victor. Designed to specification E.6/48, the HP 88 was something of a cosmopolitan aeroplane. Detail design was entrusted to General Aircraft Limited of Feltham, under their design number GAL.63, to ease the burden on the

Below *The HP 88,* VX330, *flying from Carnaby in June 1951* (via Richard Riding).

Handley Page staff involved with the full scale HP 80 (Victor), and to speed up the work a Supermarine Attacker fuselage was to be used. To assist in the joining of the 0.4 scale crescent wing to the fuselage, Swift wing roots were substituted for the normal Attacker's straight wing attachments. By the time manufacture was to begin, General Aircraft had been merged with Blackburn Aircraft Ltd and the project was moved from Feltham to Brough, Yorkshire. As a result, the design received the Blackburn designation YB 2. The fuselage was also given the Supermarine type number 521.

The aircraft, powered by a Rolls-Royce Nene turbojet, was completed in June 1951 and taken to the former wartime emergency airfield at Carnaby, whence the first flight was made on June 21 with Blackburn chief test pilot G.R.I. Parker at the controls. It was afflicted by a pitching oscillation which restricted the speed attainable and this necessitated modifications to the tailplane to effect an improvement. However, on August 26, after the aircraft had been ferried to Stansted, Essex, for further testing, it broke up in the air over the airfield killing Douglas Broomfield, Handley Page's deputy chief test pilot.

Hawker Hunter

Following the Hawker P.1081, the design for a jet fighter was refined further in Sir Sydney Camm's P.1067 scheme for the specification F.43/46. The Ministry of Supply, in March 1948, issued a new specification, F.3/48, to cover development of the P.1067 and three prototypes were ordered. Two different engines were to be employed, the Rolls-Royce Avon in the first two aircraft and, in the third, the Armstrong Siddeley Sapphire developed from the Metro-Vick F.9. Armament was to be four 30 mm Aden cannon which, with their ammunition, were in a removable pack allowing re-arming to be carrieed out within five minutes.

The first prototype was completed in June 1951 and taken to Boscombe Down for flight testing. On July 20, Squadron Leader Neville Duke made the first flight. The second prototype first flew on May 5 1952 and was the first to have the gun pack fitted. The Sapphire-powered prototype's maiden flight

The first production Hunter F 1, WT555 State Express, *being flown by Neville Duke* (British Aerospace via Bill Gunston).

followed on November 30 1952.

Onset of the Korean War prompted large production orders to be placed, with Hawker for the F Mk 1 (Avon version) and with Armstrong Whitworth for the F Mk 2 (Sapphire model), in October 1950. The first production aircraft from each of the two batches first flew on May 16 and October 14 1953 respectively. The discovery that the planned use of the main landing flaps as airbrakes was unsuccesful led to a delay in service entry while alternatives were investigated. The solution eventually found was a single large airbrake mounted under the rear fuselage. No 43 Squadron at Leuchars, Fife, was the first to be equipped with the Hunter F Mk 1 (the name bestowed in 1952) in July 1954 and later the same year F Mk 2s went to No 257 Squadron at Wattisham, Suffolk.

Development of an F Mk 3 was mooted in 1951 using an Avon engine with reheat but the sole result of this was the modification of the first prototype as a trials aircraft. The development was abandoned in favour of more powerful marks of the standard Avon and Sapphire engines. However, the performance of the modified prototype was such that an attempt on the absolute speed record was possible and this was made by Duke on September 7 1953, achieving a speed of 727.624 mph, just 12 mph faster than the previous record set in July by a North American F-86D Sabre.

The next marks to enter production were the F Mks 4 and 5 which applied extra fuel tankage in the wings and provision for underwing carriage of drop tanks or bombs to the F Mks 1 and 2 respectively. These began to join the squadrons in 1955.

The F Mk 6 brought an increase in performance with an uprated Avon engine and became the most prolific variant of the Hunter series. Mk 6s formed several renowned aerobatic teams, including No 111 Squadron's 'Black Arrows' with up to 22 aircraft and the 'Blue Diamonds' of No 92 Squadron.

The T Mk 7 brought the first change in designation of role, this being a two-seat trainer, to specification T.157D, based on the Hunter F6 with a new cockpit and forward fuselage from which the gun pack was omitted. A pair of Aden guns was, however, installed in bulges under the fuselage. The

Wingtip tanks were experimented with on Hunter F Mk 6 XG131 *early in 1956* (Hawker Aircraft Ltd via Bill Gunston).

Royal Navy also received Hunter trainers, these being similar to the RAF version but designated T Mk 8, with only one Aden gun or devoid of fixed armament.

A requirement for a ground attack fighter to replace the de Havilland Venom in tropical service resulted in the Hunter FGA Mk 9. These aircraft, converted from Mk 6s, were fitted with a tail braking parachute and had wing mounting points for four 230-gallon drop tanks or for carriage of various weapons including 1,000-lb bombs, 100-gallon napalm bombs and 2- and 3-in rockets. Also based on the F6 was the FR Mk 10, this again having the tail parachute and provision for four drop tanks, with a new nose containing three reconnaissance cameras.

Ex-RAF F Mk 4s underwent conversion to GA Mk 11s for the Royal Navy, in whose hands they were used for advanced training of pilots for the Scimitar and Sea Vixen with which the Navy was

Hunter medley: from front to back, F Mk 4 with Ecko radar nose and Fireflash missiles; F Mk 6 rebuilt as the P.1109B, the fastest Hunter of all; regular F Mk 6 with ground-attack rockets; and F Mk 6 equipped for special drop tank and 'chute trials (British Aerospace via Bill Gunston).

Latest Hunter variant is the T Mk 8M specially converted to serve as a systems trainer for future Sea Harrier pilots (British Aerospace via Bill Gunston).

re-equipping at the start of the 1960s. The designation T Mk 12 was given to a special avionics trials aircraft for use by the Royal Aircraft Establishment.

License production of several Hunter variants was carried out in the Netherlands and Belgium and standard aircraft were exported to Jordan, Iraq and Rhodesia. In addition a massive series of export marks was produced, mostly by re-building ex-RAF aircraft, and these went to Europe, the Middle and Far East, Africa and South America. (For further information on the Hunter, see Francis K. Mason's *Hawker Hunter, Biography of a thoroughbred*, also published by Patrick Stephens.)

Short Sperrin

As the Meteor and Canberra led the way in the fighter and tactical bomber roles, so the Sperrin was intended to usher in the jet heavy bomber. The specification concerned was B.14/46 which was not issued until October 1948. Design work at Short began in 1947 but a hold-up of several months during the final stages of the company's move from Rochester, Kent, to Belfast, meant a delay in the issue of detail design drawings which in turn delayed the construction of the two prototypes. Begun in 1948, the first of these, initially termed Short SA 4, was nearing completion early in 1951 and was taken by road to RAF Aldergrove for final assembly and preparation for flight trials.

On August 10 1951 Tom Brooke-Smith, Short's

The first Sperrin prototype, VX158, *fitted with Gyron engines when photographed here in September 1955* (Mike Bowyer).

chief test pilot, successfully completed the maiden flight. Unfortunately, by then, it had been decided that the type would not be ordered into production (owing to the success of the Vickers Valiant) and no further aircraft would be built, the two prototypes being assigned to an experimental role. (A third airframe had been ordered for static test only and this underwent structural load testing to destruction in 1949.) The second prototype took to the air for the first time on August 12 1952 and the following year was handed over to the Ministry of Supply for a long series of weapons trials at RAF Woodbridge.

The aircraft (named Sperrin in 1954) had no defensive gun armament but the specification had required an RCM capability, including the provision of a 'window' chute. The bomb bay was designed to carry a nuclear device weighing 10,000 lb, and was 30 ft in length and 10 ft square, this resulting in the deep fuselage and shoulder-mounted wing. A total, short range, capacity of 20,000 lb of conventional bombs was required.

The first prototype was fitted with radar for navigation and bombing trials and these were carried out from RAE Farnborough. In 1955 this aircraft was used in test-bed trials of the de Havilland Gyron turbojet, first with one Gyron in place of the port lower Rolls-Royce Avon and later with an additional Gyron in place of the starboard lower Avon. During the first flight with the two Gyrons on June 26 1956, the port outer undercarriage door fell off over the Irish Sea and was subsequently replaced with the identical item from the second prototype which, thereafter, never flew again. It was broken up in 1957. The first aircraft flew on until the cancellation of the Gyron engine early in 1957 and was scrapped at Hatfield, whence most of the testing had been carried out.

Supermarine 508, 529 and 525

The Type 508 evolved from the Type 505 project for an undercarriage-less naval fighter with a butterfly tail. This was adapted to specification N.9/47 and the first prototype, *VX133*, flew on August 31 1951 at Boscombe Down with Lieutenant Commander M.J. Lithgow at the controls. Deck landing trials were made on HMS *Eagle* in May 1952. The second prototype, *VX136*, with revised details, was redesignated Type 529 and first flew on August 29 1952. Both machines were armed with four 30 mm Aden guns in the bases of the engine air intakes.

The Type 525 was originally the third prototype of the Type 508, *VX138*, but was completed with wings of 45° sweep back at quarter chord and a conventional layout tail unit. The first flight was made on April 27 1954. Further refinement led to the production Scimitar.

Above *One of the first photographs of the Supermarine 508* VX133 *with Mike Lithgow, the test pilot, just visible in the background* (via Bill Gunston).

Right *Supermarine 529* VX136 (Mike Bowyer).

Below *The prototype 525 that led to the Scimitar* (qv), VX138, *at Boscombe Down* (Vickers-Armstrongs via Bill Gunston).

de Havilland DH 110

The de Havilland 110 design was initially conceived to specifications N.40/46 and F.44/46 for a naval all-weather fighter and an RAF night fighter respectively. Prototype orders placed in 1949 were later cut back to two machines to a new specification, F.4/48, after the naval version had been cancelled. The design was an extension of the Vampire/Venom concept and incorporated a scaled-up version of the swept-back wing flown on the DH 108 mounted on a fuselage/nacelle containing two Rolls-Royce Avons and an armament of four 30 mm Aden guns, combined with a high-mounted tailplane between twin boom/fins. Provision for two crew members was made and the nose was to contain the AI radar.

The first prototype was completed in 1951, Group Captain John Cunningham making the first flight from Hatfield on September 26. In the following year the second aircraft flew in July but during the SBAC show at Farnborough in September, tragedy struck when the first prototype disintegrated during a high-speed pass over the airfield, killing test pilot John Derry and his observer Anthony Richards and several spectators who were hit by one of the falling engines.

It was decided that the Royal Air Force would equip with the rival design to F.4/48, the Gloster Javelin, and the DH 110 lay dormant until renewed interest shown by the Royal Navy brought about an extensive re-design resulting in the Sea Vixen.

Gloster Javelin

Conceived as a competitive design against the DH 110, originally to specification F.44/46 and later F.4/48, the Javelin (then Gloster GA 5) was a large aeroplane using the delta wing planform and was destined to be the first purpose-designed all-weather jet fighter to join the RAF. Prototypes were ordered in 1949 for comparative trials with the de Havilland design. As with the rival aircraft, provision was to be made for up-to-date AI radar and an armament of four 30 mm Aden guns, but Gloster chose to power their aircraft with a pair of Metro-Vick F.9 engines (soon to become the Armstrong Siddeley Sapphire).

The first prototype was ready for flight trials by

Below *The second DH 110,* WG240, *photographed at Farnborough shortly before the tragic accident in which John Derry was killed* (Mike Bowyer).

Right *Classic photograph by Russell Adams of Javelin FAW 2/5* XA644 (Gloster Aircraft via Bill Gunston).

XA644
XA644

November 1951 and the first flight was made on the 26th of that month by Squadron Leader W.A. Waterton from the company's airfield at Moreton Valence. A re-design of the rear fuselage was found necessary to cure rudder buffet but in June 1952 elevator flutter developed during low-level trials and led to the loss of the aircraft; Waterton, who was piloting at the time, just reached Boscombe Down for an emergency landing and received the George Medal for his determined action.

However, it had been decided to develop the Gloster design for service use under the super-priority scheme and the second prototype took to the air on August 21 1952. The third aircraft, which was the first to be fitted with guns, made its maiden flight on March 7 1953. An important modification to the wing was the adoption of compound taper on the leading edge where the section outboard of the wing cannon bays was reduced in sweep back from 39.5° to 33.8°. The main purpose of this was to improve manoeuvrability at high altitude. The revised wing was first flown on the second prototype on May 28 1953, this aircraft being lost in a crash a fortnight later. However, the accident was not due to the new wing and this feature was used throughout the life of the Javelin. The fourth and fifth prototypes flew in 1954, the latter machine introducing a new cockpit canopy, and on July 22 the first production FAW Mk 1 made its maiden flight.

A further year and a half of development trials was needed before the first aircraft were handed over to the Central Fighter Establishment at West Raynham, Norfolk, in January 1956, to be followed in February by the first squadron allocation to No 46, stationed at Odiham, Hampshire. One other squadron received the initial variant before the FAW Mk 4 became the second to enter service. The other marks were as follows.

FAW Mk 2 Similar to the Mk 1 but with American-designed AI Mk 22 radar in place of the British AI Mk 17.

T Mk 3 Dual-control trainer version to specification T.118D with lengthened fuselage and heightened cockpit canopy to accommodate the instructor's raised seat behind the pupil pilot. AI radar deleted.

FAW Mk 4 Introduced all-moving horizontal tail surfaces.

FAW Mk 5 Extra 250 gallons of fuel in new wing tanks giving an increase in range.

FAW Mk 6 Similar to Mk 5 but with the AI 22 radar as in the Mk 2.

FAW Mk 7 Higher-powered Sapphire engines fitted plus armament variations, viz, only two Aden guns but, in addition, four de Havilland Firestreak

Javelin FAW 7, later converted to FAW 9 standard, at Farnborough (Mike Bowyer).

air-to-air missiles on underwing racks. Provision for fitment of four 100-gallon drop tanks for ferrying purposes.

FAW Mk 8 Final production version with, in addition to Mk 7 features, limited reheat Sapphires, American radar and wing leading edge droop.

FAW Mk 9 Mk 7s with improvements to bring performance up to Mk 8 standard, including the reheat Sapphires and drooped leading edge; British AI 17 radar, four Firestreak/four Aden armament and some aircraft fitted with fixed flight refuelling probe.

Javelins remained in service until April 1968 when the last squadron equipped with them, No 60, based at Tengah, Singapore, was disbanded.

Supermarine Swift

Following the first Type 541, which first flew in August 1951 and suffered damage in a crash landing in September, came the true prototype of the Swift incorporating detail refinements and an increase in fuel capacity giving a range theoretically superior to that of the rival Hawker Hunter. This aircraft made its maiden flight on July 18 1952. A production order had been placed in November 1950, at the same time that the prototypes were ordered, and consequently the first production aircraft, designated Swift F Mk 1, flew only five weeks later, on August 25. Armament of the initial service variant was restricted to a pair of 30 mm Aden cannon mounted in the fuselage underneath the cockpit. Actual entry into service with the RAF was destined to be severely delayed by powerplant and control problems and it was not until the spring of 1954 that a few Mk 1s joined No 56 Squadron based at Waterbeach, though even then they were subject to strict limitations on performance.

The F Mk 2, which appeared in December 1952, had a further pair of Aden guns installed, one beneath each air intake. The ammunition for these was accommodated in the wing root and to achieve this the chord at the wing root was increased. As a result further problems of control arose, the eradication of which involved modification of the wing leading edge in the form of an outboard extension. F Mk 2s joined the Mk 1s of 56 Squadron in August 1954, this being the only unit to receive fighter variants, albeit in a non-operational role and devoid of ammunition.

The Avon engine with reheat first appeared in the F Mk 3 which introduced a modified wing with the leading edge extensions and powered ailerons, a small batch of these preceeding the proposed

Supermarine 535, the prototype Swift, at Farnborough in 1950 (Mike Bowyer).

quantity production F Mk 4 variant which also featured slab horizontal tail surfaces. However, in 1955, development of the Swift fighter was cancelled.

Previously, on September 25 1953, Lieutenant Commander M.J. Lithgow—flying the prototype Mk 4, *WK198*—broke the world speed record set only 18 days earlier, in the Hunter, by Neville Duke. The attempt was flown at low level over the Libyan Desert, to benefit from the high temperature, a speed of 735.702 mph being achieved. (Only eight days later the record was again broken by the American Douglas XF4D-1 Skyray.)

Despite the fighter cancellation, work on further marks continued for use in the reconnaissance role, the tactical FR Mk 5 being developed from the F Mk 4 with a lengthened nose carrying three cameras and provision for an under-fuselage fuel tank. Armament of two Aden cannon was carried with additional provision for underwing mountings for 2,000 lb of bombs or rockets. The Mk 5 saw service with two squadrons of the 2nd Tactical Air Force in Germany. The next mark, the PR Mk 6, was to be a strategic reconnaissance aircraft with an increased wing span and no armament. Development of the Mk 6 was abandoned but the new wing was used in the final variant, designated F Mk 7, which was a small-quantity missile trials aircraft used by No 1 Guided Weapons Development Squadron at Valley, Anglesey. From March 1957 firing trials with the Fairey 'Blue Sky' (Fireflash) took place, but both aircraft and weapon were destined for cancellation with the appearance of the 1957 Defence White Paper.

Avro Vulcan

Conceived originally by Avro's technical director Roy Chadwick to meet Air Ministry specification B.35/46, the Avro 698 design was taken over after his death in the crash of the Tudor 2 prototype, by the firm's chief designer S.D. Davies. The delta wing planform was selected as it was considered to

Left *Prototype Swift F Mk 4,* WK198, *piloted over Ryde, Isle of Wight, by Mike Lithgow in August 1953* (Vickers-Armstrongs via Bill Gunston).

Below *Classic shot of a Vulcan B 2 at Farnborough in 1961* (Mike Bowyer).

provide the best answer to the specification's requirements for carrying a heavy load at high altitude and at a high cruising speed while providing good manoeuvrability. The design was ordered into the prototype construction stage in January 1948 and the first of two aircraft made its maiden flight on August 30 1952 from Woodford. The pilot on this occasion was Wing Commander R.J. Falk and the aircraft was powered by four Rolls-Royce Avon engines which had been installed in place of the planned Bristol Olympus which were not yet ready. The aircraft made an appearance at the SBAC show at Farnborough only two days later, a measure of the success of the design. A production order had been placed in June 1952. The second prototype, powered by the Olympus engines, first flew on September 3 1953 and the first production aircraft followed on February 4 1955.

The B Mk 1 was capable of carrying nuclear weapons or a conventional load of up to 21 1,000-lb bombs. No defensive armament was carried as it was considered that the aircraft could not be intercepted at the high altitudes at which it was to be operated. Before the Vulcan entered RAF service a drooped wing leading edge extension was added to delay the onset of buffeting, and this was first flown on the second prototype on October 5 1955. All the production aircraft have featured this characteristic kinked leading edge.

The first Mk 1s were issued to No 230 OCU at Waddington in February 1957 and the first operational squadron to equip was No 83, at the same base, on July 11 1957.

Development of the Avro Blue Steel rocket-powered nuclear weapon led to the appearance of the Vulcan B Mk 2 equipped to carry it. The second generation Bristol Olympus engines of greater thrust demanded a larger wing to obtain the full benefit of the power increase, and the advance in anti-aircraft missile technology demanded the inclusion of electronic equipment to combat the new threat. The second prototype was fitted with the new wing and in this guise first flew on August 31 1957. The first production B Mk 2 followed on August 19 1958. B Mk 1s were retrospectively fitted with the electronic counter-measures equipment in a new tail cone and were redesignated B Mk 1A. The Mk 2 first entered service with the pioneer

Last touchdown for B 2 XH554 *at Catterick in June 1981, before being handed over to the School of Fire Fighting there* (RAF Strike Command).

Short SB 5 WG768 *in July 1955* (Mike Bowyer).

Vulcan squadron, No 83, in July 1960 and the first Blue Steel-equipped unit was No 617 Squadron, receiving the missiles in 1962 and becoming operational in February 1963.

Trials were also carried out with two dummy Douglas Skybolt ballistic missiles, one under each wing, but the real items were never employed as it was decided that the nuclear deterrent role would be the Royal Navy's responsibility in the form of Polaris-equipped submarines. In 1966 the Vulcan force was transferred to a low-level penetration role in the face of increased radar and missile defences and has maintained this role into the 1980s. The Blue Steel equipment was phased out in 1969 as the Polaris submarines became operational. In spring 1982 several Vulcans were hastily converted to carry 1,000 lb bombs for use against Argentinian targets in the Falklands.

In 1973 a new variant of the Vulcan was introduced to service by No 27 Squadron, the SR Mk 2 being a strategic reconnaissance aircraft adapted from the B Mk 2 to replace the Victors which had been withdrawn for conversion to the tanker role.

Short SB 5

The English Electric P.1, designed to specification F.23/49, was to feature a wing of extreme sweep back which raised questions regarding the slow speed behaviour of the aircraft, for example when landing. To explore the characteristics of such a wing an experimental prototype was put out for tender, resulting in an order to Short of Belfast to Ministry of Supply specification ER.100.

The SB 5, as it was called, was taken to Boscombe Down where, on December 2 1952, Tom Brooke-Smith made the first flight. The aircraft was designed to test various configurations of wing and tail combination and to begin with a wing of 50° sweep back with a delta tailplane atop the fin was tried. The next stage was to increase the sweep back to 60° and in this form it started flying in July 1953. In January 1954 the third variation—60° wing with drooped inboard leading edge and horizontal tail surfaces at the bottom of the fuselage—appeared. This was the nearest layout to that of the P.1 which first flew the following August.

Four years later the SB 5 was returned to Belfast for extensive alteration. The Rolls-Royce Derwent originally installed was replaced with a Bristol Orpheus engine of smaller size and greater power, an ejector seat and new instrumentation were installed and a wing of 69° sweep back fitted. The low mounted tail surfaces were maintained. In September 1960 the aircraft was transported to RAE Bedford whence the first flight was made on October 18. After completing its final research programme at Bedford, the aircraft was handed over to the Empire Test Pilots' School at Farnborough where it was used to give pilots experience in low speed handling of slender aircraft. It 1968 it was retired and began a tour of duty as a ground instructional airframe, passing through Finningley and Topcliffe and eventually, in 1974, becoming an exhibit at the Aerospace Museum at RAF Cosford.

Handley Page Victor

Designed and developed in parallel with the Avro Vulcan to specification B.35/46, the Handley Page team, under the leadership of R.S. Stafford, produced an aircraft of totally different appearance from the Avro submission. Using a conventional fuselage-and-wings layout, the design nevertheless broke new ground by using the crescent wing in which the angle of sweep back was reduced in three stages from the fuselage outwards to the tip. This planform was adopted after studies of German wartime research material had shown that it offered the optimum performance at high subsonic speeds. The horizontal tail surfaces were mounted on top of the sharply swept fin. The powerplant chosen was to be four Armstrong Siddeley Sapphires mounted in pairs in each wing root.

Two prototypes of the HP 80 were ordered in April 1948 and construction commenced in 1951, the first machine being ready for transport by road from Radlett to RAE Boscombe Down by May 1952. The aircraft was re-erected but the first flight was delayed by technical problems coupled with a fire in the rear fuselage during systems tests. Eventually, after a period of poor weather during the week before Christmas when the first taxiing trials were made, the maiden flight was achieved on the 24th with H.G. Hazelden as pilot. Trials continued successfully until July 1954 when, flying from Cranfield, the aircraft was destroyed in a crash during low-level airspeed calibration flights. The second prototype was completed as quickly as possible to fill the gap and first flew on September 11.

Concurrently with the Vulcan order in June 1952, a batch of Victor production aircraft was ordered, the first of these flying on February 1 1956. Receiving the designation B Mk 1, delivery to No 232 OCU at Gaydon commenced on November 28 1957 to be followed by the first operational unit, No

A Victor K1 tanker of 57 Squadron at Marham in 1974 (MoD).

10 Squadron at Cottesmore, being declared operational on April 15 1958. The Mk 1 was able to carry a varied weapon load of up to 78,000 lb, considerably greater than the Vulcan. The B Mk 1A was distinguished by the incorporation of electronic counter-measures equipment, some of which was carried in a new tail cone.

Development of the Victor in the flight refuelling tanker role was accelerated by the grounding of the Valiant tankers in 1964 which left the RAF without any refuelling aircraft. The first of them, the B (K)Mk 1As, were B Mk 1As fitted with two hose-and-drogue pods, one under each wing, and these entered service in May 1965. There followed a more fully equipped version with a third fuselage-mounted hose-and-drogue and these first entered service in January 1966. Converted from B Mk 1s and B Mk 1As, they were designated K Mk 1 and K Mk 1A respectively.

As with the Vulcan B Mk 2, so a more advanced Victor was developed with the Blue Steel-carrying capability. In addition a wing of increased span was used and the powerplant changed from Sapphires to four Rolls-Royce Conways. The prototype Mk 2 first flew on February 20 1959 but, on August 20, it disappeared over the sea off the South Wales coast. A massive fleet of trawlers and other vessels carried out an extensive search and eventually located the wreckage under 400 ft of water. Ultimately some three-quarters of the aircraft (by weight) was recovered in several hundred thousand fragments and these were 'reconstructed' by the RAE at Farnborough to ascertain the cause of the accident. It was finally discovered that the breaking off of the starboard pitot head had caused the aircraft to enter a dive which was beyond the control of the crew. A simple modification was all that was needed to prevent the situation recurring.

Production Mk 2s entered operational service in February 1962, the first squadron to be equipped

The first Victor K2 on entry into service with 232 OCU at Marham (MoD).

being No 139 at Wittering. The operational Blue Steel-carrying Victors were called B Mk 2R and included several refinements, the most noticeable being the wing trailing edge aerodynamic bodies which delayed transonic drag rise and contained 'window' (anti-radar foil strips). No 543 Squadron began receiving, in May 1965, the SR Mk 2 strategic reconnaissance variant which featured increased fuel load and various camera equipments for day or night recce missions.

The Victor bombers had been deployed in the low-level role, in common with the Valiant and Vulcan, since 1964 and remained in service until 1968 when they were relinquished for conversion to the tanker role, thus leaving the Vulcan as the sole long-range medium bomber in service. In addition the SR Mk 2s were phased out in 1975, also to be converted into tankers, the first of which, designated K Mk 2, had flown on March 1 1972. They started their service lives in May 1974 at Marham, initially with No 232 OCU, equipping No 57 Squadron, the first of the operational squadrons to receive them in 1975.

Short Sherpa

Specification B.35/46, which resulted in the Vulcan and Victor, elicited from Short a design for an aeroplane based on the Sperrin but with a new swept wing and fin and no horizontal tail surfaces. This wing was of fundamentally new concept in that it was designed to overcome the problems associated with control at high subsonic speeds—buffet of trailing edge control surfaces and flexure of the wing structure causing a dangerous variation in incidence. To overcome these problems the tips of the wings formed the control surfaces, acting as elevators or ailerons depending on whether they were moved in unison or in opposition and the wing structure contained the spar/torsion box much further aft than normal to eliminate the twisting action which would allow incidence variation. This so-called 'aero-isoclinic' wing was developed by Short's chief designer David Keith-Lucas in conjunction with Professor G.T.R. Hill but, unfortunately for the company, their bomber design submission was not accepted.

However, in order to prove the theory in practice, Short decided to go ahead and build a one-third scale flying model of the bomber at their own expense. (Therefore the Sherpa may be seen to be akin to the Avro 707 and Handley Page HP 88 which flew the wings of the Vulcan and Victor respectively.)

Top and above *The Short Sherpa* G-14-1 *with Tom Brook-Smith at the controls* (via Richard Riding).

Initially a glider, called the SB 1, was built but this, first flown by Tom Brooke-Smith in July 1951, was involved in a crash landing in October after loss of control when passing through the slipstream of the towing aircraft. As a result Brooke-Smith received injuries which resulted in six months' recuperation and a determination that glider trials should not be continued. Therefore a revised airframe, using the wings and tail of the glider, was built and was fitted with a pair of Turboméca Palas turbojets to allow powered trials to be carried out. In revised form, now designated SB 4 and named Sherpa, Tom Brooke-Smith made the maiden flight on October 4 1953 from Aldergrove. The aeroplane proved to handle well and trials continued throughout 1954 but although other designs featuring the 'aero-isoclinic' wing were drawn up by Short (including a rival to the future Blackburn Buccaneer), no further aircraft were actually constructed.

The Sherpa was subsequently passed on to the College of Aeronautics at Cranfield from where further flights were made until 1964 when it became a ground test airframe at the Bristol College of Advanced Technology. Two years later it became a museum piece at the Skyfame Museum at Staverton.

Miles Sparrowjet

On July 13 1957 the famous King's Cup Air Race was won for the first time by a jet-powered aircraft. Even more remarkable was the fact that the winner had first competed in the 1935 Race, albeit in a substantially different form.

The prototype Miles M.5 Sparrowhawk, *G-ADNL*, powered by a de Havilland Gipsy Major engine, had been placed eleventh in the 1935 Race owing to severe handicapping, though its average speed of 172 mph was only four miles per hour slower than the winner. The Sparrowhawk competed again in the 1936, 1937 and 1949 King's Cup Races and during that time changed hands twice. In 1950 it was acquired by Fred Dunkerley and handed over to F.G. Miles at Redhill for rebuilding as a jet racer.

Three years' work, during which the company moved to Shoreham, resulted in an aeroplane of considerably altered appearance which was given the designation M.77 and name Sparrowjet. Removal of the de Havilland engine from the nose allowed the pilot to occupy a new cockpit forward of the wing, beneath a large clear canopy. The resulting increase in fuselage length necessitated revised tail surfaces, including finlets at the ends of the tailplane, to maintain directional stability. The new powerplant of two Turboméca Palas jet engines was installed in a pair of small nacelles, one in each wing root close to the fuselage. The undercarriage was considerably shorter than before as there was no need for ground clearance for a propeller.

The 'new' aircraft, still with the registration *G-ADNL*, was flown for the first time from Shoreham by George Miles on December 14 1953. Its first race win was the 1956 SBAC Challenge Cup at Yeadon with a speed of 197 mph. In the King's Cup Race of 1957 Fred Dunkerley achieved an average speed of 228 mph.

The Sparrowjet's subsequent history was sad. After being handed over to BEA, its life ended at RAF Upavon in July 1964 when it was destroyed in a hangar fire.

Miles Sparrowjet G-ADNL *carrying also the race number* 92 *after its victory in the 1957 King's Cup Race* (via Richard Riding).

Hunting Jet Provost and BAC Strikemaster

Development of the P.84 Jet Provost began in the spring of 1951, as a private venture by Hunting Percival Aircraft Ltd from the piston-engined Percival P.56 Provost trainer. The new aircraft was to utilise the wing and tail components of its predecessor combined with a new fuselage containing an Armstrong Siddeley Viper turbojet engine and with a side-by-side two-seat cockpit forward. In 1953 the Ministry of Supply expressed their interest in the new trainer by placing an order for nine machines to be designated T Mk 1 for use in a series of tests. The first of these made its maiden flight from Luton on June 16 1954 and deliveries began the following year to No 2 FTS, Hullavington, Wiltshire, for service suitability trials. They were the first purpose-built jet trainers to join a Royal Air Force unit.

A tenth Mk 1 aircraft was built with refinements including a shorter undercarriage and, with a more powerful Viper engine, this, together with four similar demonstration aircraft, received the designation T Mk 2. From this variant stemmed the production T Mk 3, fitted with ejector seats and wingtip tanks, which entered service in June 1959. The T Mk 4 was similar to the Mk 3 but had a further uprated engine giving a notable increase in performance. Overseas demonstrations led to several purchases, there being two marks involved, both featuring armament. Two 0.303-in machine-guns were mounted in the fuselage beneath the air intakes and underwing mountings were provided

Jet Provosts from Cranwell's The Poachers *aerobatic display team in May 1976* (MoD).

Red, white and black Strikemaster demonstrator (Darby Childerhouse, BAe).

for up to 360 lb of rockets or bombs. Otherwise similar to the T Mks 3 and 4, the overseas versions were the T Mks 51 and 52 respectively, examples of the former going to Ceylon, the Sudan and Kuwait and of the latter to Iraq, Venezuela, the Sudan and South Yemen. (The Yemen received aircraft converted from ex-RAF machines.)

The next model in the Jet Provost series—the Hunting H.145 and later BAC 145—was a considerable advance over the Mk 4 and had a redesigned forward fuselage and pressurised cockpit for high altitude training. The wing structure was revised to allow greater fuel capacity and in production aircraft, T Mk 5s, the wing is the strengthened unit designed for the Strikemaster. The prototype Mk 5, converted from one of the last two production T Mk 4s, made its maiden flight from Warton on February 28 1967. The Central Flying School was the first unit to receive the new mark in September 1969. A variant of the Mk 5—the T Mk 55 fitted with an armament as the earlier overseas versions—was sold to the Sudan.

The BAC 166, converted from the last production Mk 4, was a one-off test bed for the Viper 20 engine which flew on March 16 1965 and was subsequently re-converted to become the second prototype T Mk 5.

The ultimate design of the series was the BAC 167, later named Strikemaster, developed as a ground attack and trainer aircraft for the export market. Being similar in appearance to the T Mk 5, the new design has the Rolls-Royce Viper 20 engine, two 7.62 mm FN machine-guns in the fuselage beneath the air intakes and a strengthened wing allowing the carriage of a weapon load of up to 3,000 lb at eight positions. The first flight was made on October 26 1967 from Warton. Given designations in the Mk 80 series, sales have been made to Saudi Arabia, the Yemen, Oman, Kuwait, Singapore, Kenya, New Zealand and Ecuador.

English Electric P.1 and Lightning

Conceived by W.E.W. Petter, originally to Ministry of Supply specification ER.103 for a supersonic research aircraft, the English Electric P.1 was ordered into the initial construction stage as two prototypes and one static test airframe to specification F.23/49 with a view to development as a high performance operational interceptor. The design, having a number of advanced features, notably the wing of 60° sweep back, was extensively tested in model form in a specially built (by English Electric) transonic wind tunnel to prove the feasibility of the chosen layout. In addition Short was contracted to design and build an experimental machine to prove the handling qualities of the wing (the SB 5). The tests proved that the aircraft would work and construction went ahead as planned, the first machine being ready for flying in 1954.

Transported by road to the A&AEE airfield at Boscombe Down, the P.1 flown by Wing Commander R.P. Beaumont made its first flight on August 4. On the third flight a speed in excess of Mach 1.0 was achieved. The second prototype joined the programme on July 18 1955 and was equipped with several new features including an additional fuel tank faired into the underside of the fuselage, a fin of increased area, outboard wing leading edge extensions and an armament of two 30 mm Aden guns mounted alongside the cockpit.

In 1954 (design work by then had been taken over by F.W. Page after Petter had left the company in 1950) a contract had been placed for a developed version to be designated P.1B, three aircraft to be built featuring new equipment more akin to that needed for an operational aircraft. The original powerplant of two Armstrong Siddeley Sapphires with primitive afterburning was to be replaced with a pair of Rolls-Royce Avons of greater sophistication; the nose air intake was redesigned to a circular section containing a pressurised centre body housing an AI 23 radar; and the fuselage was increased in size to allow for the bigger engines. A choice of armament was provided for in removable ventral packs; either a pair of de Havilland 'Blue Jay' (Firestreak) air-to-air missiles on external pylons with their electronics inside the pack or a pair

Right *Lightning F3s of 'Treble One' armed with Red Top missiles at Akrotiri, Cyprus, in January 1974* (RAF).

Below right *Camouflaged Lightning of 92 Squadron at Gutersloh* (RAFG).

Below *The first P.1,* WG760, *later restyled P.1A* (British Aerospace via Bill Gunston).

K

Lightning T4 of 5 Squadron armed with Firestreak missiles in March 1973 (MoD).

of 30 mm Aden cannon plus ammunition or two boxes of 24 2-in Folding Fin Aircraft Rockets.

The first P.1B took to the air on April 4 1957. A production order for 50 aircraft, to include 20 pre-production examples for weapon and equipment testing, was placed in November 1956 and the first of these flew on April 3 1958. The name Lightning was given in October 1958 and the following month, on November 25, the first P.1B, flown by Beamont, attained a speed in excess of Mach 2.0, the first time a British aircraft had achieved this.

The first full production standard Lightning F Mk 1 flew for the first time on October 29 1959 and deliveries began at the end of June 1960 to No 74 Squadron at Coltishall. Owing to lack of endurance because of low fuel capacity, a flight refuelling probe was added under the port wing, aircraft with this facility being redesignated F Mk 1A.

In 1961 the Lightning F Mk 2 appeared, externally similar in appearance to the Mk 1 but with Avon engines with fully variable reheat, deliveries of this version starting in 1963 and equipping two squadrons. These aircraft were subsequently uprated by the addition of the cambered wing and belly fuel tank of the Mk 6, becoming F Mk 2As.

An improved radar and compatibility with the Red Top missile were features of the F Mk 3, which also had a further increase in engine power and a new square-tipped fin of greater area to improve directional stability at high speeds. Entry into service began in 1964 with No 74 Squadron again, then based at Leuchars. The next two marks were trainer variants of the F Mks 1A and 3, with two-seat side-by-side cockpits, designated T Mks 4 and 5 respectively. The former went into service in 1962 followed in 1965 by the Mk 5.

The final major development of the Lightning, the F Mk 6, first flew on April 17 1964. The wing originally flown on the second P.1, with reduced outboard leading edge sweep back, plus increased camber and inset ailerons, was used and a larger ventral fuel tank replaced that fitted in previous marks. Pylon-mounted auxiliary fuel tanks can be mounted above the wings to increase the ferry range. Again No 74 Squadron pioneered the service acceptance of the latest Lightning, receiving aircraft in September 1966.

Saudi Arabia and Kuwait have been the only other countries to purchase and operate Lightnings, placing orders in 1966. A developed version of the F Mk 6, called the F Mk 53, was supplied to both countries, these aircraft having an underwing weapons carrying capability for bombs or rockets, and the T Mk 55 trainer version was also built. In addition the Royal Saudi Air Force received an interim batch of ex-RAF aircraft designated T Mk 52 (ex-F2s) and T Mk 54 (ex-T4s). (For further information on the early history of the Lightning, see Bill Gunston's *Fighters of the Fifties*, also published by Patrick Stephens.)

Folland Midge and Gnat

Having initiated the P.1 design for English Electric and left that company in 1950, W.E.W. Petter became managing director of Folland Aircraft Ltd of Hamble, Hampshire. There he turned his mind to development of a lightweight fighter of comparatively simple design which would be considerably less expensive than other types then under development (including the P.1). Such an aeroplane, he reasoned, might be of considerable interest to countries without the facilities to operate increasingly sophisticated types or lacking funds to buy them.

In 1952 two designs were formulated, the Fo.140/1 to be powered by a Bristol Saturn engine and the Fo.140/2 with a Rolls-Royce Derwent. However, before anything came of these, the Saturn engine was cancelled and the Derwent was considered unsuitable. The following year a revised design was drawn up around a new Bristol engine then under development. A prototype was started which would prove the airframe using an Armstrong Siddeley Viper engine, while the Bristol powerplant was awaited for the definitive version. The Viper-powered aircraft, Fo.139 and named Midge, was quickly completed and taken, assembled, by road to Boscombe Down on the morning of August 11 1954. The first flight was made that afternoon in the hands of Squadron Leader E.A. Tennant and three further flights were completed before it left, in the evening, for the company's Flight Development Unit at Chilbolton, Hampshire. Testing progressed smoothly, allowing the Midge to appear at the Farnborough show in September where a great deal of interest was shown in it. As a result several countries sent pilots to fly the aircraft, including Jordan, New Zealand and India. Unfortunately disaster struck on September 26 1955 when it crashed on take-off from Chilbolton during a flight by a Swiss pilot who was killed in the accident.

However, on July 18 the definitive aircraft had flown. This, the Fo.141 Gnat with its Bristol Orpheus engine, was equally successful and an order for six was placed by the Ministry of Supply for evaluation flying. These aircraft had an armament of two 30 mm Aden cannon, one mounted in the outer wall of each air intake. The fighter was not selected for RAF use but the sixth aircraft of the MoS batch was sent out to India as a result of an order from that country which subsequently involved license production by Hindustan Aircraft Ltd and later still the development of their own Ajeet. Other countries also bought Gnat fighters (Fo.145s): two were acquired by Yugoslavia and 12 went to Finland.

Although the RAF did not require the fighter, in 1957 an order was placed for a batch of Fo.144 two-seaters for use as advanced trainers. The first of

The Folland Fo. 139 Midge flying over the Hampshire countryside (Folland Aircraft via Mike Bowyer).

Facing page *For most people, the diminutive Gnat will always be associated with the* Red Arrows, *the RAF's premier display team whose red-painted aircraft achieved world-wide fame before being replaced by the modern Hawk. Here they are seen in 1974—the year in which the team won the coveted Air League Founder's Medal* (MoD).

these made its maiden flight on August 31 1959 and deliveries began in February 1962 to the Central Flying School at Little Rissington, Gloucestershire. During a long career in RAF service, which lasted into the late 1970s, the Gnat was immortalised by its performance as an aerobatic aeroplane. In 1964, No 4 FTS at Valley, Anglesey, formed a team of five yellow-painted Gnats known as the 'Yellowjacks' and in 1965 CFS formed the nine-strong 'Red Arrows' team, with whom the Gnat served for 15 years.

Fairey FD 2

In 1947 the Ministry of Supply began to consider how Britain could make up lost ground in the field of supersonic aircraft development and, after discussion with individual companies, issued a specification for a research aircraft with supersonic capabilities. This specification, ER.103, was to result in the two most important (technologically) jet aeroplanes to be designed before the mass amalgamations of the aircraft industry: the English Electric P.1 (from which came the Lightning) and the Fairey Delta Two.

The Fairey design marked a major advance for the company since their main work previously had been on the long line of naval aircraft including the Swordfish and Firefly and most recently the Gannet, which was under intensive development owing to the waging of the Korean War. This delayed the design work on the FD 2 until 1952 and manufacture began only towards the end of that year. To achieve as clean a design as possible meant the inclusion of several advanced and novel features amongst which were the main undercarriage units which retracted forward, rotating through 90° to lie flat in the wing, and the so-called 'droop-snoot' in which the entire cockpit and forward fuselage could be depressed 10° to give an improved forward view to the pilot when landing. Leading the design team were Fairey's chief designer H.E. Chaplin and chief engineer R.L. Lickley.

The first of two aircraft, *WG774*, was completed in mid-1954 and taken to Boscombe Down for flight trials, making the first flight on October 6 piloted by Peter Twiss. Thereafter the speed attained on each flight was increased toward Mach 1.0 until, on the 14th flight on November 17, engine failure occurred at 30,000 ft when the aircraft was 30 miles away from Boscombe Down. Twiss nursed the invaluable machine back to base and landed at over 250 mph. The aircraft left the runway and sustained damage to the wing which necessitated an eight-month break in flying while repairs and adjustments were made.

In August 1955 the Delta Two was ready to

Fairey Delta Two, WG774 (Fairey via Mike Bowyer).

resume flying and steadily proceeded through its programme, making the first supersonic flight on October 28. The Rolls-Royce Avon engine was fitted with an afterburner which had not yet been used and it was clear that the power as yet untapped would enable a speed in excess of the current absolute speed record to be easily attained. Fairey therefore set about the business of preparing for an attempt, and it was eventually arranged to use a course on the South Coast between Chichester and RN Air Station Ford. The aircraft would be flying at 38,000 ft and this meant the use of very sensitive measuring equipment on the ground to record the speeds on the runs made over the course. Also, precise calibration of the Delta Two's altimeter was imperative to ensure that the altitude variation did not exceed that allowed by the FAI regulations. The attempts began on March 8 1956 and the runs which were to count as the record submission were made on the 10th. The average speed achieved was 1,132.136 mph, an improvement of more than 300 mph over the previous American-held record set by a North American F-100C Super Sabre.

The second FD 2 first flew on February 15 1956 and exceeded Mach 1.0 during the 25-minute flight. In October, this machine was flown out to Cazaux,

in the South of France, to carry out supersonic trials at levels down to 3,500 ft. Both aircraft subsequently carried out a comprehensive flight programme in Britain based at RAE Bedford and in 1961 the record-breaking aircraft was handed over to the British Aircraft Corporation for conversion to the BAC 221.

de Havilland Sea Vixen

Deck landing touch-and-go trials with the second prototype DH 110 were carried out aboard HMS *Albion* in the autumn of 1954, after modifications had been effected since the fatal accident at Farnborough. These included saw-tooth leading edges and a slab tailplane in place of the variable incidence tailplane and conventional elevator. The aircraft first flew in this revised form on June 11 1954. Meanwhile, work was started on a 'navalised' interim aircraft called the Sea Vixen Mk 20X which started flying on June 20 1955. By then an order had been placed by the Ministry of Supply for three pre-production development aircraft and a batch to full production standard to specification N.139P. The Mk 20X carried out full arrested-landing trials on HMS *Ark Royal* beginning on April 5 1956, and the first of the pre-production aircraft was flying by the end of March 1957. This and the second machine took part in a series of trials at Boscombe Down and RAE Bedford, and at sea in *Ark Royal* and *Centaur*.

The Flying Trials Unit, No 700 Squadron's 'Y' Flight, received its first Sea Vixen FAW Mk 1s in November 1958 for working up and the first front line squadron to be equipped with the type was No 892, which formed at Yeovilton in July 1959. The Sea Vixen was the first British interceptor to dispense with guns in favour of missile armament, having provision for two nose-mounted retractable packs containing 14 2-in rockets in each and four underwing-mounted de Havilland Firestreak air-to-air missiles. Alternatively, an underwing load of four 500-lb or two 1,000-lb bombs could be applied for strike missions. Drop tanks could also be carried to extend range for ferrying or reconnaissance missions, and Flight Refuelling Mk 20 drogue refuelling pods could be carried for in-flight transfer of fuel.

Left *Superb air-to-air view of a Sea Vixen Mk 2 at the time of its entry into Royal Navy service in 1962* (Hawker Siddeley via Terry Gander).

Below *The Sea Vixen in its natural environment leaving* Eagle*'s catapult while a plane-guard Wessex hovers alongside* (MoD via Terry Gander).

Vixen Mark 1 in 1963 (Mike Bowyer).

In 1963 the second production variant appeared, the FAW Mk 2, which had increased internal fuel capacity in forward extensions of the tail booms, and this mark was able to carry Red Top air-to-air or Bullpup air-to-surface missiles. No 899 Squadron was the first to receive the Mk 2 in December 1963, embarking in HMS *Eagle* the following year. In 1965 the ship was involved in the Beira patrol and the Sea Vixens were used to keep an eye on the movements of tankers trying to supply fuel to Rhodesia after the UDI. No 899 was destined to be the last squadron to relinquish its Sea Vixens when *Eagle* paid-off for the last time in 1972 but several machines continued to be used by the RAE in various research programmes and a batch was converted for use as pilotless drones for missile development trials with the designation D Mk 3.

Supermarine Scimitar

The Scimitar evolved from the Type 525 to the definitive specification N.113D and production proceeded to specification N.113P. The new aircraft featured several advancements over its progenitors, notably an area-ruled fuselage and a system of flap blowing using high-pressure air ducted from the engine compressors and expelled through a slot ahead of the flap hinges to delay boundary layer flow breakaway. This was of value in particular in reducing the approach speed for carrier landings and had been tried on the Type 525. In addition, the wings had outboard leading edge extensions and boundary layer fences and the tailplane was mounted at 10° anhedral, all of which helped to counteract the 'pitch-up' phenomenon encountered at high speed.

Lieutenant Commander M.J. Lithgow made the Type 544's maiden flight from Boscombe Down on January 19 1956 and carrier trials began in April on board HMS *Ark Royal.* The first production aircraft flew on January 11 1957. The production aircraft, designated F Mk 1, which had a fixed armament of four 30 mm Aden cannon (in the position originated in the Type 508), was destined to be operated in the low-level strike role in which a variety of underwing-mounted weapons could be carried. On four wing positions these included 500 or 1,000 lb bombs, Bullpup or Sidewinder missiles, pod- or rack-

Supermarine 544 being flown by Mike Lithgow in 1956 (via Mike Bowyer).

Production Scimitar photographed at the SBAC show at Farnborough in September 1959 (Mike Bowyer).

mounted 2-in and 3-in rocket projectiles or nuclear weapons.

No 700X Flight (later No 700 Squadron) based at Ford, Sussex, received Scimitars in August 1957 for intensive flying trials and the first front line unit to equip was No 803 Squadron at Lossiemouth in June 1958, subsequently embarking in HMS *Victorious* in September. In September 1964, Scimitars of 'B' Flight, No 800 Squadron, embarked in HMS *Eagle*, served as flight refuelling tankers, maintaining this role for two years; the rest of the squadron was flying Buccaneers which had superseded the Scimitar in the strike role.

Short SC 1

In 1953 the Rolls-Royce Thrust Measuring Rig, better known as the 'Flying Bedstead', began to explore the possibilities of vertical take-off using jet engines. The rig employed a pair of Nene engines, horizontally mounted, whose thrust was directed downwards at the machine's centre giving lift, lateral control being provided by compressed air ducted from the engine compressors to four 'puffer-pipes' mounted at 90° to each other on outriggers. Following successful trials with the 'Bedstead', Rolls-Royce initiated the design of a small turbojet specifically for vertical mounting, under the direction of Dr A.A. Griffith. The Ministry of Supply supported this research effort and issued a specification, ER.143, for an aeroplane to explore the new engine's capabilities.

Short started design work in 1954 on a small delta-winged machine, design sequence number SC 1, to be powered by five of the Rolls Royce RB.108 engines, one mounted at a 30° angle in the rear fuselage and exhausting horizontally for forward flight and four, centrally grouped, in the vertical position for lift. Construction of two prototypes began in 1955 and the first of these, with the propulsion engine only, began taxiing trials at the company's Sydenham airfield on December 17 1956. In the following March, the aircraft was shipped to Southampton and taken by road to Boscombe Down for flying trials. The first conventional flight was made on April 2 1957. In Belfast, the second aircraft with its lift engines installed commenced ground running in September but it was not until May 23 1958 that a tethered hover was accomplished by test pilot Tom Brooke-Smith. The lift engines were mounted in pairs which could be tilted fore and aft through 35° to assist in transition to forward flight or to decelerate for vertical landing.

On October 25 1958 Brooke-Smith made the first

Below and bottom *The Short SC 1 at Farnborough the following year* (Mike Bowyer).

free hovering flight and the following month landed the aircraft away from the point of take-off. Gradually the flight 'envelope' was explored more fully until, on a flight from RAE Bedford on April 6 1960, the first full transition from horizontal flight to vertical and back again was accomplished.

Continual improvements were applied to the two SC 1s, engine power was increased and new autostabilisation systems were employed. During a test flight involving the second aircraft in an autostabilisation programme on October 2 1963, an RAE test pilot, J.R. Green, was killed when at low level a gyro failure caused the aircraft to flip over and crash to earth. However, it was repaired and returned for further trials, eventually joining its companion at RAE Bedford for extensive VTOL research.

Miles Student

The Miles M.100 was developed as a private venture competitor to the Hunting Jet Provost for use as a jet primary trainer. Construction was begun at Shoreham in 1955 and the distinctive aeroplane which took shape, with its dorsal-mounted jet engine and twin fins, was completed early in 1957. The first flight on May 14 1957 was made by George Miles and in September it was demonstrated at the Farnborough Air Show. Unfortunately for the company the need to build the aircraft without official support meant that development had to take second place to money-making orders and the delay thus imposed meant that the Jet Provost was accepted into RAF service before the Student's first flight.

However, the one aircraft built was to have a career of its own. The original Turboméca Marboré 2 engine was replaced by a more powerful Marboré 6F and thus powered it first flew on April 22 1964 prior to taking part in gunnery trials based on Boscombe Down. Two underwing pods, each containing one 0.303-in machine-gun and its ammunition, were carried, and firing trials were made at the Larkhill ranges. Alternative underwing loads included rocket projectiles and light bombs.

A naval deck-landing trainer was proposed and it was hoped that the Student might gain acceptance as a civil trainer for airline pilots, but neither of these schemes was realised. Expression of interest by South Africa, for license building in that country for use by the South African Air Force, came to nothing when a government embargo was placed on

Miles Student at its Farnborough debut in 1957 (Mike Bowyer).

arms shipments. The sole aircraft had been dismantled and packed for the journey when the blow fell and was thereafter put into storage at Shoreham in October 1965.

Resurrection came in 1974 when a Ministry of Defence funded scheme for research into aircraft noise reduction resulted in refurbishment of the Miles Student to take part in the trials. Modifications for the trials were made at the Hawker Siddeley Flight Test Centre at Dunsfold and the aircraft was flown over the Hatfield runway for noise measurements to be made. The trials were successfully completed over a six-month period, the Student acquitting itself well, with the exception of its engine (the original Marboré 2) which was somewhat low on power. It was returned to Shoreham on completion of its programme in October 1974.

Saunders-Roe SR 53

In the early 1950s aircraft designers began to explore the possibilities of a mixed-power interceptor in which two engines would be installed, one a small turbojet for cruising flight and the other a high performance rocket motor with which rapid take-off and climb to altitude could be made, then being shut down until an enemy was located and opened up again for a speedy interception.

In 1952 a Ministry of Supply specification, F.124D, was issued to the leading manufacturers calling for an undercarriage-less aircraft which would be ramp launched, returning to earth on a landing skid and being picked up by a special

The first prototype SR 53 on its second flight on May 27 1957 (British Hovercraft Corporation Ltd via Bill Gunston).

recovery vehicle. The small Cowes firm of Saunders-Roe had not been considered for issue of a copy of F.124D, although Maurice Brennan, a member of their design team, had been making detailed studies of a mixed-power aircraft as a company private venture. As a result they requested a copy of the specification in order to submit a design based on their findings. Brennan prepared a submission to meet the requirement, adding a rider for a more conventional design which, actually, would have been of superior performance.

With little delay the second design was accepted and a new specification, F.138D, composed to cover it. Three prototypes were ordered and detail design started in 1953. The SR 53 was planned to attain a speed of Mach 2.2 at altitudes in excess of 45,000 ft and carry an armament of two wingtip-mounted Blue Jay (Firestreak) missiles. However, there was insufficient room to install an AI radar, so concurrently a second design for a larger aircraft to be fitted with an AI Mk 23 set was drawn up, this having a further covering specification issued, F.177D. Nine prototypes were to be built of the SR 177 including three each of specific Royal Navy and Royal Air Force versions.

The first of the SR 53 prototypes started flying in 1957 from A&AEE Boscombe Down, the initial flight being made by Squadron Leader J.S. Booth on May 16. Flight testing was proceeding smoothly and construction of the SR 177 prototypes was in hand, although a shadow was spreading over the scene in the form of the 1957 Defence White Paper in which it was envisaged that manned aircraft would be phased out of RAF service in favour of missiles. This resulted in the cancellation of the RAF SR 177. Development continued on the naval version and it was expected that the Luftwaffe and Kriegsmarine would place a large order. Then the entire programme was cancelled.

Test flying of the SR 53 prototypes continued (the second machine had joined the first in December) though they were then in effect only research aircraft, when the final tragedy in the story came on June 5 1958. Squadron Leader Booth, while taking off from Boscombe Down, ran off the runway after an aborted take-off and hit a concrete approach light post, a section of which broke off and fell on to the cockpit, killing him.

Blackburn Buccaneer

In June 1952 the Admiralty issued the NA.39 requirement for a carrier-borne two-seat bomber aircraft capable of carrying a nuclear weapon at low level (to avoid enemy radar surveillance) at a high speed. This requirement was translated into the Ministry of Supply specification M.148T in August for distribution to the aircraft manufacturers.

The Blackburn company, based at Brough,

Buccaneer S1 XN928 *in 1962* (Mike Bowyer).

Buccaneer being launched from Ark Royal *in 1975* (MoD).

Yorkshire, tendered the B.103, designed under the leadership of B.P. Laight, and this was accepted in July 1955. The gap of nearly three years between the issue of M.148T and acceptance was due to the extreme technical advance which the requirement represented. The B.103 featured an area-ruled fuselage to minimise the drag induced at high speed, which allowed the use of a fuel-efficient powerplant giving a respectable load-carrying range. The small wing, desirable for low-level manoeuvrability and carrier stowage, was achieved by using the boundary layer control flap blowing system already applied to the Supermarine Scimitar.

An order was placed for 20 pre-production aircraft, there being no prototype as such, each of which would be used to explore a specific area of the type's performance. The first of these, which was essentially an aerodynamic test vehicle, made its maiden flight from Thurleigh (RAE Bedford) on April 30 1958 piloted by Lieutenant Commander D. Whitehead with B.J. Watson as observer. A production order was placed in September 1958 for 40 aircraft, later to be designated Buccaneer S Mk 1.

Deck landing trials, using two of the pre-production batch, began in January 1960 aboard HMS *Victorious*, which during her recent extensive modernisation had had her flight deck and arrestor gear strengthened to operate the new 'heavyweight'. Even before the first production standard aircraft had flown, No 700Z Flight, the Intensive Flying Trials Unit, had been formed at Lossiemouth, Morayshire, with six of the pre-production NA.39s. The first S.1 flew on January 23 1962 and aircraft began joining the initial operational unit, No 801 Squadron, in July. The squadron formed at Lossiemouth for working up, subsequently joining HMS *Ark Royal* in February 1963.

While the S1 was entering service the uprated S Mk 2 was under development. In this variant the Gyron Junior engines were replaced by the more powerful Rolls-Royce Spey which, giving 30 per cent more thrust, used less fuel and allowed take-off at a greater all-up weight. The Spey engines were installed in a converted Mk 1 which first flew on May 17 1963 and the first production S2 took to the air on June 6 1964. No 700B Flight began trials with the new Buccaneer in March 1965 and No 801 Squadron became operational in October, again at Lossiemouth, first going to sea with *Victorious* in June 1966.

Developed from the S Mk 2 was the S Mk 50 for the South African Air Force. Amongst the modifications made to meet the requirements of that service was the incorporation of a retractable pack under the rear fuselage containing a Bristol Siddeley BS 605 rocket motor for take-off assistance in high temperatures. The aircraft were delivered to South Africa in 1965.

In 1968 Buccaneers were ordered for the RAF to fill the gap left by cancellation of the TSR 2 and F-111K. In addition to a batch to be built specifically for the RAF, the service would receive the naval aircraft released with the phasing out of

Buccaneer S.2B of 16 Squadron, Laarbruch, in 1974 (RAFG).

the aircraft carrier. The land-based Buccaneers have an extensive range of weapons at their disposal up to a total of 16,000 lb. This may include six 1,000-lb or 12 500-lb bombs on the four wing pylons; a 4,000-lb load on the rotating bomb bay door; wing-mounted rocket pods or Martel air-to-surface missiles. The bomb bay may also be used for a reconnaissance camera pack or a fuel tank containing an extra 440 gallons. Distinctions of mark are the S Mk 2A and 2B in which the latter is specifically equipped to carry the Martel missile. The remaining Royal Navy machines were re-designated S Mk 2C and 2D respectively.

The first new-build aircraft for the RAF flew in January 1969 and in October of that year No 12 Squadron became the first unit to form on the type, based at Honington, Suffolk, and tasked with a maritime strike role.

An S2B inside one of the hangars at Laarbruch (RAFG).

Hawker P.1127 and Kestrel

At the time of the 'missiles only' Defence White Paper of 1957, which seemed to spell the end of military aircraft development by British companies, a project was under way at Bristol Engines involving a single powerplant of completely new concept. Based on an idea of the French Engineer Michel Wibault, it involved a ducted fan engine in which all thrust would be emitted through a set of rotatable nozzles, a forward pair taking cold air from the compressor and a rear pair forming the outlets from the bifurcated tail pipe. The nozzles, when in the vertical position would direct the thrust downwards thus providing lift, and rotated through 90° aft would impart forward flight, the lift then being provided by the aircraft's wings. (During transition, stability would be afforded by compressed air cascades fore and aft and at the wing tips.) This eliminated the need for separate engines, as in the Short SC 1, for lift and forward flight.

The Hawker company, under the leadership of Sir Sydney Camm, began work on a potential fighter/ground-attack aircraft using the engine, as a possible candidate for current NATO requirements. Owing to the official policy the company undertook the job as a private venture while the engine development proceeded with financial assistance from the NATO Mutual Weapons Development Programme. America was interested in the project from the start and provided assistance in the form of aerodynamic research facilities at NASA centres; in addition Hawker chief test pilots A.W. Bedford and H. Merewether gained jet-lift experience flying the Bell X-14 research aircraft.

Development of the P.1127 proceeded apace during 1959; in June the following year the Ministry of Supply belatedly stepped in with a contract for two prototypes to 'research' aircraft specification ER.204D and followed up with an

Left *P.1127* XP831 *at Farnborough in 1962* (Mike Bowyer).

Below *Five Kestrels of the Tri-Partite Evaluation Squadron at West Raynham in 1956* (MAP).

order for four further development prototypes five months later. By then the first Bristol BE.53 engine (later named Pegasus) had been delivered and in October 1960 the aircraft was ready to begin trials. On the 21st of that month, at Dunsfold, Bill Bedford performed the first tethered hovering flight over a special grid which eliminated the ground effect caused by the downwash of the hot exhaust gases. A series of taxiing trials was then carried out to be followed on November 19 by the first untethered hover over the grid. After further hovering tests the P.1127 was taken to the RAE airfield at Bedford where the first conventional take-off was made on March 13 1961, again with Bill Bedford at the controls. Progress towards the change from hover to forward flight and back again culminated on September 12 in the first successful transition.

The second prototype first flew from Dunsfold in a conventional take-off on July 7 1961 and was initially involved in low-speed handling trials. On September 22 the first machine made a successful vertical lift-off from solid ground.

In 1962 an important stage in the development of the aircraft was reached when a collaborative decision involving Britain, West Germany and America resulted in the ordering of nine production aircraft for a tri-national evaluation. All six prototypes had flown by the end of 1962 but a year later only three remained: the second machine had been abandoned in flight after one of the cold air nozzles had been lost, the third P.1127 was badly damaged in an emergency landing at Tangmere and the first aircraft had the misfortune to suffer a nozzle control failure during a demonstration at the 1963 Paris Air Show, being badly damaged in the ensuing crash. The three survivors continued to carry out trials pending the appearance of the first production aircraft, that event occurring on March 7 1964 when the first flight was made by Bill Bedford.

The Tri-partite Evaluation Squadron was formed at the Central Fighter Establishment, West Raynham, on October 15 and extensive trials were made by pilots from the RAF, Luftwaffe, US Army, Navy and Air Force. The aircraft, which had been designated Kestrel F (GA)Mk 1, were (with the exception of one lost in an accident) shipped to America for further trials after nearly a year with the TES during which some 938 flights had been made. The Kestrels were designated XV-6A in US service.

Handley Page HP 115

Co-ordinated research into the development of a supersonic airliner was begun in 1956 under the auspices of the Supersonic Transport Aircraft Committee and by the end of the decade designs featuring slender delta wings and flying at speeds in the region of Mach 2.2 were being favoured. However, very little was known of the handling characteristics such a wing might display at low speeds and it was decided therefore to draw up a specification for a research aircraft to explore those qualities. Originally consideration was given to a glider but it was decided that a powered aircraft would be more versatile and therefore specification

The HP 115 photographed in September 1962 (Mike Bowyer).

ER.197D was composed and issued to Handley Page.

The resultant machine, design number HP 115, was built during 1960. A wing with 74° 42′ leading edge sweep back was featured with a low thrust Bristol Siddeley Viper turbojet mounted above it aft beneath a sharply-swept fin and rudder. The cockpit was in a nacelle at the apex of the wing, the pilot's shoulders being level with its leading edge. There was a fixed tricycle undercarriage.

The HP 115 was taken by road to RAE Bedford where an extended series of taxiing trials was made before venturing into the air. On August 17 1961 Squadron Leader J.M. Henderson took the machine up for the first time for a flight lasting just over half an hour. There were no serious problems encountered and the aircraft was shown at the SBAC show at Farnborough the following month.

Low-speed handling trials occupied the next four years and produced a mass of data towards development of the airliner which would later be named Concorde. One aspect of the trials was the exploration of the wing vortices, these being made visible by smoke generators mounted on the leading edge aft of the cockpit, the results being recorded by a cine camera mounted in a pod near the top of the fin. Further trials occupied the aircraft until 1974 when it was retired, having reached the end of its designed flying life of 500 hours. It went to the historic aircraft collection at RAF Colerne and was later transferred to the museum at RAF Cosford.

Hawker Siddeley Trident

Trident began life as the Airco DH 121, a short-haul airliner specifically suited to the needs of BEA. The BEA specification was originally issued in 1956 and resulted in submissions from the leading manufacturers, foremost of which were those from Bristol, Avro and de Havilland. A condition of the award of the contract for development of the selected design was that it should be undertaken by a consortium. For that reason the Aircraft Manufacturing Company Limited was formed in 1958 by the de Havilland, Hunting and Fairey concerns. (This resurrected the name of the firm from which de Havilland had evolved in 1920.) In 1959 the design was adjusted to meet a BEA demand that the aircraft would be larger than current passenger traffic required and in the following year the merging of de Havilland with the Hawker Siddeley group led to the winding up of Airco. Thereafter development was solely the responsibility of the former de Havilland designers and the product would become the Hawker Siddeley HS 121.

Several aspects of the aircraft were new in British designs, in particular the triple engine installation at the rear which was intended to keep the passenger

Right *British Airways Trident 3 over central London* (BA). **Below** *An earlier model at Farnborough in 1964* (Mike Bowyer).

cabin free of noise. It also left the wing (of 35° sweep back at quarter chord) aerodynamically 'clean'. The HS 121 was the first airliner to have provision for full automatic landing incorporated at the outset with a triple hydraulic system operating in parallel, each driven from one of the engines, in which failure of one system would be automatically over-ridden by the two remaining. Appropriately the name Trident was bestowed in September 1960.

Owing to the reputation of the de Havilland company no prototype as such was to be built, a production batch of 24 being ordered 'off the drawing board'. Construction took place at Hatfield and the first machine was rolled out in August 1961. Fitted with ground running engines in the port and centre positions, systems and taxiing trials were carried out prior to delivery of three flight-cleared engines. These were installed in time for a first flight on January 9 1962 with John Cunningham as first pilot assisted by Peter Buggé. The first five Tridents were used for development trials and in December 1963 the sixth machine was delivered to BEA for crew training. The seventh, delivered the following February, made the first passenger-carrying flight to Copenhagen on March 11.

An improved version of the Trident 1, the 1E, with increased fuel capacity, greater wingspan and more powerful engines, was sold to several overseas airlines including those of Kuwait, Iraq and Pakistan. Developed from the 1E for BEA was the Trident 2E which included still more fuel, uprated engines and structural strengthening, the first of this model flying on July 27 1967. The Chinese airline, CAAC, placed orders for the 2E beginning in 1971

and eventually running to a total of 33. They also ordered two of the ultimate development, the Super 3B.

The Trident 3B was a considerable advance over its predecessors with a fuselage stretch allowing an increase from 140 seats to 180 (maximum), a slightly larger wing and a fourth boost engine beneath the rudder which increases thrust for take-off by 15 per cent, allowing operation from more restricted airfields. The 3B flew for the first time on December 11 1969 and entered service with BEA in 1971. Further increase in fuel capacity and higher all-up weight were features of the Super 3B of which only the two Chinese aircraft have been built, the initial machine making its maiden flight on July 9 1975.

Bristol Type 188

As supersonic aircraft were being developed capable of progressively greater speeds (up to the region of Mach 2.0) there seemed good reason to build a research aircraft to investigate the behaviour of airframe materials at the high temperatures encountered at those speeds. This phenomenon, known as kinetic heating, is a result of the friction between the aircraft's skin and the air passing over it which becomes greater as the speed is increased. (For the same reason space capsules returning to the Earth's atmosphere at great speed require a heat shield to prevent their being burnt up.)

In the mid-1950s an extremely advanced bomber was specified which was to be capable of a cruising speed of Mach 2.5 at an altitude of 60,000 ft. The Avro submission was chosen for development and, as with previous advanced designs, scaled-down research aircraft were to be built to prove various aspects of the new bomber's configuration and performance. One of these was ordered from Bristol Aircraft and given the type number 188. Its specific purpose was to test the design of the engine nacelles at high speeds. With the appearance of the 1957 Defence White Paper (envisaging the end of manned aircraft), the Avro bomber was cancelled but the Bristol Type 188 was reprieved by specification ER.134 for the purpose of general research into kinetic heating.

Two aircraft and one static test airframe were

The first Type 188, XF923, *being flown by Godfrey Auty, Bristol's chief test pilot, from Boscombe Down in 1962* (Bristol Aircraft via Mike Bowyer).

British Airways VC 10 at Heathrow (BA).

ordered and these were built at Filton. Long delays in their completion were caused by the use of stainless steel in their airframes, which demanded totally new techniques of construction, and by the use of an untried engine, the de Havilland Gyron Junior. Eventually the first prototype was completed in early 1962 and made its maiden flight on April 14, with the second aircraft flying over a year later on April 29 1963.

Unfortunately, the aircraft was not a success largely due to the extremely fuel-thirsty engines which did not allow sufficient duration for the prolonged high-speed flying necessary for the research programme. In 1964 the project was terminated and the aircraft disposed of. One of them (the second) found its way to the weapon ranges at Foulness for use as a ground target, whence it was eventually rescued for restoration as a museum exhibit. It is at present at RAF Cosford.

Vickers VC 10

In 1957 BOAC announced that a new jet airliner, to be built by Vickers, would be acquired to operate on their medium range routes as a replacement for the turboprop Britannia. The design, based on studies by Vickers on swept wing jet developments of their Vanguard, was subsequently altered into a long-range airliner powered by four Rolls-Royce Conways and using systems technology developed for the cancelled V.1000 military transport. In 1958 an order was placed by BOAC for 35 of these VC 10s and work was commenced at the company's Weybridge factory in January 1959.

The first aircraft, type number 1100, made its maiden flight from Weybridge on June 29 1962, to be followed on November 8 by the first production standard type 1101. The ninth aircraft built was the first to be delivered to BOAC, on April 22 1964, and this inaugurated fare-paying operations on West African routes a week later.

Meanwhile, in 1960, BAC (of which Vickers had become a part) had announced the Super VC 10 with longer fuselage, uprated engines and greater fuel capacity. This version could carry up to 187 passengers, 36 more than in the standard aircraft, and BOAC cut their original order for 35 to 12, ordering 30 super VC 10s instead. This quantity was later reduced to 17. The first type 1151 took to the air on May 7 1964 and the third began scheduled service flying on April 1 1965. A small number of both types has been operated by overseas airlines, Ghana receiving new-built VC 10s and East African Super VC 10s; in addition to these Gulf Air, Laker, Middle East and Nigeria have operated secondhand machines and others, such as Air Ceylon have leased aircraft from BOAC.

The Royal Air Force accepted VC 10s into service in answer to the 1960 specification C.239 for a long-range supplement to Transport Command's

Comets. The military variant, type 1106, designated C Mk 1, was dimensionally similar to the standard VC 10 but featured the increased fuel capacity and uprated engines of the Super. They also have a side-loading freight door, a strengthened floor, rear-facing passenger seats, flight refuelling probe and an auxiliary Bristol Siddeley Artouste engine mounted in the tail cone. The first flight of a C Mk 1 took place on November 26 1965 and deliveries to the sole squadron to operate the type, No 10 based at Fairford (and later at Brize Norton), commenced on July 7 1966. Each of the 14 aircraft received the name of a recipient of the Victoria Cross.

In 1975, cutbacks in defence expenditure resulted in the disposal of four of the C Mk 1s but work on a new VC 10 variant began in 1979 after the announcement that the current Victor tanker fleet was to be expanded by converting nine ex-airline aircraft. Five standard VC 10s will become K Mk 2s and four Supers will be K Mk 3s.

An RAF VC 10 XV109 *being serviced at Akrotiri en route to the Far East* (RAF).

Hawker Siddeley HS 125

With development of the DH 125 Jet Dragon, de Havilland introduced the executive jet to Britain's aircraft industry. With a crew of two and accommodation for six passengers, it has been a popular choice with many companies operating their own fleets of business aircraft and has had success in North America, competing against the wide range of indigenous designs in the class.

A wooden mock-up was built in 1961 to display the layout of the new design which was intended as a successor to the piston-engined Dove. Construction of two prototypes at Hatfield and of 30 Series 1 production aircraft at Chester was begun. The first prototype was flown for the first time on August 13 1962 and the second followed in December, these then being used in certification trials. The name Jet Dragon was shortlived and since 1964 the series has been known simply as the HS 125 (BAe 125 since 1978). The first Series 1 aircraft made its maiden flight on February 12 1963, the production version featuring a three-foot greater wing span and an extra foot of fuselage length over the prototypes.

The six main series variations have introduced stage-by-stage improvement of internal arrangements and uprated engines. The Series 600 represented the first major advance with an increase in length allowing a more spacious cabin for eight executive-style seats or up to 14 standard seats, improved systems and an enhanced power-to-weight ratio giving a more efficient performance. New Garrett AiResearch turbofans introduced in the Series 700 of 1976 ensured that the HS 125 remained competitive, giving, as they do, a reduction in fuel consumption and noise levels and the ability to operate to and from 'hot and high' airfields.

Military development of the design has given the Royal Air Force its first tailor-made jet-powered navigation trainer. First flown in December 1964, the RAF aircraft was named Dominie and the 20 aircraft built constituted the Series 2 production version of the HS 125. The Dominie T Mk 1 entered service with No 1 Air Navigation School at Stradishall in December 1965. Two further military designations have been given to service versions of the Series 400 and 600, these becoming the CC Mk 1 and 2 respectively, operated in the communications role by No 32 Squadron at Northolt, in particular for the transport of government ministers, high-ranking officers and other VIPs.

HS 125 G-BFAN *showing its paces at the 1976 SBAC Show* (Mike Bowyer).

Hunting H.126

Among the schemes explored for shortening the take-off run of aircraft was the so-called 'jet-flap' principle, in which a proportion of the efflux from a turbojet engine is ducted into the wing and exhausted through a narrow slit in the trailing edge. This layer of hot gas when deflected downwards by the flaps can provide a considerable increase in lift by forming, in effect, an increased area of flap surface.

To test this theory a research aircraft was called for in specification ER.189D and from the Luton-based Hunting Aircraft came the H.126 to meet the requirements. The aircraft was powered by a Bristol Siddeley Orpheus engine, the larger part of its efflux being ducted to the trailing edge slit and the remainder exiting via nozzles mounted low in the fuselage behind the wing. Both sources provide thrust. Pitch, yaw and roll control was administered by jet nozzles in the wing tips and tail (similar to the puffer-pipes in the VTO P.1127 series), in addition to elevators and ailerons in the normal positions. As this was not a high performance aircraft a fixed tricycle landing gear was adopted to avoid the cost and complication of a retractable undercarriage.

The H.126 was taken to the RAE Bedford to carry out its research programme and made its first flight there on March 26 1963. Its programme of trials occupied the next four years and in 1969 the aircraft was shipped to America for wind tunnel tests at the NASA Ames Research Laboratory in California. It was returned to England in 1970 and put into storage pending disposal. In 1974 it became part of the collection at the Aerospace Museum, RAF Cosford, expanding the experimental aircraft contingent already on display.

The Hunting H.126 photographed in 1962 (BAC via Mike Bowyer).

BAC One-Eleven

Hunting Aircraft produced a design study in 1956 for a short-range, medium-capacity airliner to be powered by two Bristol Siddeley turbojets pod-mounted at the sides of the rear fuselage. Although a registration *(G-APOH)* was allotted for a prototype of this H.107 design, the aircraft was not built. When Hunting was absorbed into BAC in 1960, the airliner design was reworked into one slightly larger with Rolls-Royce Spey engines and a passenger capacity of 65 instead of the 48 planned for the H.107. With a planned range of 600 miles, the BAC 111, as it was then called, was developed by the former Hunting and Vickers design staffs at Luton and Weybridge. In March 1961 it was announced that an instruction to proceed had been received for a batch of 20 aircraft and, in May, that ten had been ordered by British United Airways.

Construction was undertaken at Hurn (Bournemouth) and the first machine was rolled out early in August 1963, the first flight being made on the 20th by BAC test pilots G.R. Bryce and M.J. Lithgow. The following month it was flown to Wisley to begin the development programme and had made 52 flights by October 22 when it was lost due to deep stall. The One-Eleven project pilot, Mike Lithgow, and six other occupants were killed. When the cause of the accident had been discovered from the flight recorders the information was disseminated throughout the industry in order that a repetition of the accident should not occur. As a remedy the leading edge of the wing was modified and the elevator control linkage adapted to give ample warning if an excessive angle of attack was imminent.

British Airways One-Eleven G-AVMH *at Berlin-Tempelhof. This aircraft was part of the European Division's fleet based at Manchester* (BA).

The second prototype was too advanced for modification, so after first flight on December 19 1963, it was restricted to systems trials. The fourth One-Eleven received the full modification and took to the air on April 1 1964. Full C of A clearance was issued on April 6 1965 and BUA's first One-Eleven scheduled flight was made on the 9th on the Gatwick to Genoa route.

The initial production version was the Series 200 and in common with other modern civil aircraft improvements in design and performance resulted in new series numbers being issued. By 1980 the Series 670 was the ultimate One-Eleven to appear. Uprated engines and greater range were features of the 300, the 400 added American equipment specifically for that country's requirements and the Series 500 had an increased wing span and stretched fuselage to increase passenger capacity to a total of 119 (the previous production aircraft all had a capacity of 89 passengers). Combination of the new wing of the Series 500 with the shorter fuselage of the 400 resulted in the Series 475, three examples of which have entered military service with the Sultanate of Oman Air Force.

An order placed in 1967 by the Romanian state airline, Tarom, has led to the setting up of a One-Eleven production line in that country on which, initially, aircraft are assembled from parts fabricated in Britain. The plan is that ultimately full responsibility for production will be phased in by the mid-1980s.

BAC 221

At the same time that specification ER.197D was issued to Handley Page for development of their HP 115 low-speed slender delta research aircraft, a second specification, ER.193D, was issued to the Bristol company for a further research aircraft. This was to be a rebuild of the 1956 World Speed Record-breaking Fairey Delta 2 with a new wing of ogival delta shape; which was the planform chosen for the supersonic airliner also in hand at Bristol. The 'new' aeroplane would be used for both high- and low-speed research and for the latter regime in particular the original Fairey-designed 'droop snoot' would be of major importance for landing at a high angle of attack. For the same reason the modified undercarriage included a very long nose leg. The fuselage was lengthened to allow for the longer root chord of the new wing and to avoid breaking the smooth curve of the leading edge, the engine air intakes were positioned below it, the ducts rising to form upper surface bulges in the wing root fairing.

Given the Bristol type number 221 this aircraft was, in effect, the last design of that company as a separate entity to be built, as they became part of the British Aircraft Corporation during 1960. The BAC 221 made its maiden flight at Filton on May 1 1964 and thereafter underwent a period of

The 'droop-snoot' 221 in September 1964 (Mike Bowyer).

development test flying, followed by a modification programme in which the fin was increased in height to improve directional stability, an anti-spin parachute was added and equipment refinements were made. Just over two years after first flight, on May 20 1966, the aircraft was delivered to RAE Bedford where it joined in the slender delta research already in progress with the HP 115, its flying programme continuing until June 1973.

The BAC 221 was subsequently acquired by the Museum of Flight at East Fortune near Edinburgh and has recently been moved to Yeovilton for display alongside Concorde 002.

BAC TSR 2

As early as 1952 designs had been invited, to Ministry of Supply specification B.126T, for a low-level strategic bomber, but the idea had to be abandoned owing to the non-existence of the advanced technology that would be needed for such an aircraft. Then, in 1955, the RAF began to explore the possibilities of a successor to the Canberra. At that time the Royal Navy had just accepted the Blackburn submission to meet their NA.39 requirement (the Buccaneer) and it was hoped that this aircraft might be adapted to suit the RAF's need. However, the RAF was determined that a totally new aircraft was the answer and began to formulate a requirement for a low-level supersonic aeroplane.

In the aftermath of the 1957 Defence White Paper the only project that sustained advancement was the Canberra replacement, and an RAF General Operational Requirement—GOR.339—was issued, via the Ministry of Supply, to the industry for initial development. In September 1957 a meeting was held in London at which MoS representatives and the heads of the leading aircraft companies were present to discuss the future of the industry and, in particular, its relationship with the GOR.339 project. The main outcome of the meeting was that the contract for the aircraft would be awarded to a group of companies working together and that this would be a prelude to the streamlining of the industry by amalgamations. The submissions were to be in by the end of January 1958.

The choice for development fell to the English Electric P.17A and that company would be joined with Vickers to work on the detailed design which was to be to the revised Operational Requirement OR.343 and to contract specification RB.192D. The gestation period of the new aeroplane—to be known as TSR 2 (for Tactical Strike Reconnaissance)—was prolonged by many factors, amongst which was the need to develop totally new systems

XR219, *the one and only TSR.2 to fly, during its first flight from Boscombe Down with Roland Beamont at the controls* (MAP).

and equipment for navigation, low-level terrain avoidance and accurate delivery of weapons.

Controls and strength of the basic structure to withstand the turbulent conditions which would be encountered at supersonic speed at low level posed further problems and the proposed engines, new afterburning versions of the Olympus, were dogged with difficulties to begin with. However, as the problems were surmounted it became clear that the aircraft, which was considerably in advance of anything that had gone before, would be a success and in 1963 TSR 2 was offered to the Royal Australian Air Force who were looking for a replacement for their Canberras.

While the industry was struggling, not only with TSR 2 but with the amalgamations that had been thrust upon it, the entire project became a political pawn on the grounds of cost. As the date for a general election in the autumn of 1964 loomed, the Labour party, then in opposition, played on the economic aspect in order to sway public opinion (on what was by nature a highly secret project) in their favour. If they were returned to government they would cancel outright. Adverse comment led the Australians to order the F-111 from the USA instead of the British product.

On September 27 1964, at Boscombe Down, Roland Beamont, pilot, with D. Bowen as navigator, successfully took the first aircraft on its maiden flight. Initial teething troubles with the engine and undercarriage were overcome and its qualities were being explored in the first stages of the development programme when, in the first Budget speech by the new Labour Chancellor of the Exchequer, Callaghan, in April 1965, the TSR 2 project was cancelled in its entirety.

The one aircraft which had flown and the second which was only days from joining the programme, together with five airframes in an advanced state of construction at Weybridge, were disposed of either by immediate destruction or by allocation for use as targets on the gunnery ranges at Shoeburyness. (The gutted airframes of two machines were preserved, one of which can now be seen at Duxford.) All jigs were destroyed. In their place a batch of 50 General Dynamics F-111K swing-wing aircraft was ordered from America and these, in their turn, were cancelled in 1968.

In October 1969 the Buccaneer joined the Royal Air Force.

Unusual angle on TSR.2 (BAC via Mike Bowyer).

Hawker Siddeley Harrier

While the development and trials of the Kestrel were in progress, the British aviation industry went through the upheaval of cancellation of its most advanced designs, amongst them the TSR 2 and P.1154. The latter was a supersonic V/STOL machine developed from the P.1127 series for entry in the NATO Basic Military Requirement 3 competition, with plans for separate versions for the RAF and RN. However, a survivor of this period was a redesigned Kestrel incorporating a much more powerful model of Pegasus engine with airframe refinements to suit. Under the design direction of J.W. Fozard the new machine, to be named Harrier, proceeded successfully to first flight on August 31 1966. This was the first of six pre-production machines, the first definitive Harrier GR Mk 1 taking to the air on December 28 1967.

Armament includes a pair of 30 mm Aden cannon in two detachable under-fuselage pods and various combinations of stores, including bombs (up to 5 × 1,000-lb), Sidewinder missiles and Matra rocket pods on four underwing and one fuselage pylon mountings. In addition, extra underwing fuel tanks and a five-camera reconnaissance pod on the fuselage pylon may be carried.

Contemporary with the Mk 1 was the T Mk 2 two-seat trainer with lengthened forward fuselage to allow the inclusion of a second (rear) cockpit for the instructor. To maintain the centre of gravity position, an extension of the tail boom was added to allow ballast to be carried, also to extend the tail 'puffer pipe' aft to counteract the increased fuselage length.

Further increases in engine thrust brought mark revisions, first to GR Mk 1A and then GR Mk 3; similarly the Mk 2 became the T Mk 4. A noticeable addition to the GR Mk 3 was the laser ranger and marked target seeker in a bulbous nose mounting.

Entry into service with No 1 Squadron at RAF Wittering began on April 1 1969, three further squadrons forming in the next three years, all in RAF Germany. Subsequently, No 20 Squadron relinquished its Harriers, these being used to expand the other two units.

Continued interest by America in the Harrier led to an order being placed in 1969 for the US Marine Corps. These aircraft, designated AV-8A and

Harrier GR Mk 3s of 3 Squadron from Wildenrath during an exercise in June 1974 (MoD).

Above *The two-seat Harrier demonstrator,* G-VTOL, *at Lee-on-Solent in October 1974* (HMS Daedalus).

Left *In its 'hide' in the Borken area of Germany during an exercise in 1976, a Harrier GR 3 of 4 Squadron fitted with the laser-ranging and target-seeking nose and rear warning receivers on the tail fin* (RAFG).

TAV-8A (in single-seat and trainer form respectively), introduced the 'viffing' manoeuvre to the combat capability of the type, in which the nozzles are vectored in forward flight (VIFF) allowing sudden unexpected course changes to be used to confuse and/or avoid an enemy. A combined research project between Hawker Siddeley and McDonnell Douglas brought about an advanced design, known as AV-16A, which was not in fact proceeded with, but development of the basic Harrier has resulted in the AV-8B, with super-critical section wing of greater span and area and other improvements relating to engine performance.

Uncertainty over the future of the AV-8B led to a British Aerospace (part of which Hawker Siddeley had become) scheme for a British-developed advanced Harrier with 'big wing' and lift enhancing devices, such as leading edge root extensions (LERXs) for improved performance at high angles of attack and better turning capability. The Spanish Navy has so far been the only other recipient of Harriers, these being supplied via the United States where specific modifications were made to machines built in Britain; in Spanish service the aircraft is called Matador.

Hawker Siddeley Nimrod

As the threat posed by ballistic missile-carrying nuclear submarines increased, the Air Staff began to formulate requirements for a replacement for the piston engined Avro Shackleton. In 1958 they issued ASR.381 for a multi-capability aeroplane whose prime role would be hunting and killing submarines but which could also serve as a search and rescue aircraft, general reconnaissance machine and as a troop carrier in emergency.

A wide range of studies was forthcoming, varying from all-new advanced swing-wing aircraft to adaptations of existing airliners, with either turbo-prop or turbojet powerplants. It was thought that the turbo-prop would offer advantages in terms of greater endurance because of its more economic use of fuel, but by 1964 the most favoured design was that from Hawker Siddeley proposing a Comet derivative powered by four Rolls-Royce Spey turbofans. To achieve the required range and time on station it was planned that the aircraft would fly its patrol with two engines shut down to save fuel, using all four to take off and reach its search area and again to return to base.

Based on the Comet 4C, the new HS 801's most distinctive visible variation was the much deeper fuselage brought about by adding the long weapons bay to the underside, this being neatly faired into the lines of the basic fuselage fore and aft. The nose section of this fairing formed the radome for the ASV 21 radar scanner. To counteract the increased

Protecting Britain's North Sea oil rigs is one of the Nimrod's most important tasks. Here is XV246 *off the coast of Aberdeen in 1975* (MoD).

A Nimrod of 206 Squadron from Kinloss takes off from Gibraltar during a NATO exercise in 1976 (MoD).

'keel' of the fuselage a dorsal extension was added to the standard Comet style fin, to the top of which was attached a faired pod containing the ESM aerial. At the extreme tail was the 'sting' type MAD boom.

Construction of two prototypes, converted from a pair of spare Comet 4C airframes, began in 1965 and following these was a production run of 38 aircraft. The first prototype was taken up for the first time on May 23 1967 by John Cunningham, from the factory airfield at Chester (formerly de Havilland) and flown to Woodford to start its flight test programme. The first production aircraft, now named Nimrod, flew in June 1968 and there began an intensive programme of trials to evaluate and prove the systems and avionics. By October 1969 the first MR Mk 1 was delivered to RAF St Mawgan to start training with 236 OTU and the first front line unit to receive the type was No 201 Squadron, at Kinloss, in the summer of 1970. In addition to the maritime aircraft, three Nimrods designated R Mk 1 were delivered to No 51 Squadron for service as electronic reconnaissance aircraft.

The development and installation of more advanced electronic equipment, bringing about a re-designation to MR Mk 2, has, as might be expected, considerably enhanced Nimrod's capabilities. Weapons which both versions may carry cover the entire range of modern anti-submarine equipment, from homing torpedoes, mines and depth charges to sonobuoys, all of which are carried in the fuselage bay. (Small sonobuoys can be launched from special launchers in the aft fuselage compartment.) In addition, underwing racks for up to four air-to-surface missiles can be fitted.

Rolled out in April 1980 and first flying on July 16 was the first of an entirely new variant, the Nimrod AEW Mk 3. These early warning aircraft are converted from Mk 1s released by No 203 Squadron when it disbanded after the final withdrawal from Malta. Fitted with bulbous radomes at the nose and tail containing the pulsed-Doppler scanners, the Mk 3 will be the most advanced aircraft in its class in the world. The layout was chosen to allow an uninterrupted 360° sweep, each scanner covering 180° and synchronised to form the complete 'picture'. Planned for entry into service in 1983, these Nimrods will replace the last of the Shackletons which have been serving in the AEW role.

BAC/Aérospatiale Concorde

Serious design studies for the development of a supersonic airliner were begun in Britain in 1955 and the following year the Supersonic Transport Aircraft Committee (STAC) was set up to co-ordinate the research effort. In France similar projects were started in 1957 with the Nord, Sud and Dassault concerns each exploring individual streams of research.

In 1960 steps were taken in both countries towards the selection of a specific design layout from the mass of detailed design data produced up to then. In Britain the favoured scheme was the Bristol 198, the ultimate in a long series of studies bearing that type number, which featured for the first time the wing and fuselage conformation of the Concorde as finally designed, though a powerplant of six engines was then planned. The French design was produced as a joint venture between Sud and Dassault. The following year an agreement was reached in principle on the combined development of a single aircraft based upon the further refined Bristol (BAC) 223 and Sud Aviation Super Caravelle. On November 29 1962 the official agreement for development and production of an Anglo-French supersonic transport was signed in London.

To begin with, two separate designs were worked on, one for medium and the other for long-range use, but by 1964 all effort was concentrated on the long-range version. Construction of two prototypes was started in April 1965, the first at Toulouse and the second at Filton. The French aircraft made its maiden flight on March 2 1969 and the British prototype, *002*, followed it into the air on April 9 with Brian Trubshaw at the controls.

Two pre-production aircraft followed, again one from each partner, and these incorporated refinements as a result of the test programme carried out by the prototypes. A major difference was the lengthening of the fuselage and rearrangement of the passenger cabin to include seating for up to 144 (although normal seating capacity is 128). The second, French-built, pre-production Concorde, *02*, was representative of the production machine and had a new drag-reducing fuselage tail cone added which contained extra fuel and an emergency power unit. Its maiden flight was made on January 10 1973.

The first four, of 16, production aircraft were used for development trials and it fell to the fifth and sixth to inaugurate scheduled flying. Concorde *205*, in Air France colours, flew from Paris to Rio de Janiero via Dakar and Concorde *206*, in British Airways livery, flew on the London to Bahrein route, both flights being made on May 24 1976.

While representing a magnificent technical achievement, Concorde has suffered from criticism of an environmental and economical nature and this has resulted in a drastically curtailed production run and limited service application, particularly in America. No airlines, other than the national carriers of the manufacturing countries, have bought them, although Braniff and Air Singapore crews fly sectors of the routes flown in their territories.

Concorde during a low-speed pass at Farnborough (British Aerospace via Raymond Baxter).

Sepecat Jaguar

Convergence of two national streams of development, in France and Britain, for an advanced trainer, with secondary tactical ability, was effected in 1964 when the first joint operational requirement was issued. In 1965 the Breguet Br 121 design was chosen as the basis for detailed collaborative development, in which a 50/50 sharing of responsibility was given to Breguet and the British Aircraft Corporation. Similarly, Rolls-Royce and Turboméca got together to produce an engine, subsequently named Adour, specifically for the new aeroplane. The name Jaguar was announced in the same year and in May and June 1966, two new companies were formed to look after the development of airframe and engine respectively. The former, based in Paris, named Sepecat (standing for Société Européenne de Production de l'avion d'Ecole de Combat et d'Appui Tactique) and the latter, London-based, called Rolls-Royce Turboméca Ltd.

By 1967 the tactical strike capability was becoming of greater importance and a series of five variants was decided upon. Three of these were for French service, the Jaguar E (two-seat trainer), A (single-seat tactical strike) and M (single-seat carrier-borne tactical strike), and two for the Royal Air Force, the Jaguar B (two-seat trainer) and S (single-seat tactical strike). The apparent duplication of roles concealed the fact that the British aircraft were to have more advanced equipment, for example a Martin-Baker Mk 9 zero/zero ejector seat with rocket boost and more sophisticated navigational and weapon aiming equipment. In addition to these five variants was an export model, to be based on the S airframe, termed Jaguar K.

The first aircraft to fly was the French *E-01* prototype, which took off from Villacoublay on September 8 1968, piloted by Bernard Witt. The British *S-06* prototype made its maiden flight from Warton on October 12 1969. In 1970 the strike aircraft was given pride of place in the RAF order, quantities of this version being increased from the 90 decided upon in 1967 to 165 and trainers dropping from 110 to 35 to maintain the total of 200 machines.

The Armée de l'Air introduced Jaguar to service in June 1973 and the first GR Mk 1 was delivered to the Royal Air Force in September, joining the Conversion Team at Lossiemouth. At the same

The second production Jaguar GR Mk 1, XX109, *over the Wiltshire countryside* (A&AEE).

A Jaguar T Mk 2 deploys its brake parachute on landing (BAC).

base the first squadron, No 54, began receiving its aircraft in March 1974, moving south to Coltishall, Norfolk, in August as a fully operational unit.

Armament of the British Jaguar includes a pair of 30 mm Aden cannon mounted in the lower fuselage and provision at five hardpoints (one fuselage and four underwing) for up to 10,000 lb of weapons, including bombs (conventional or cluster), Martel air-to-surface missiles and Matra pod-mounted rockets plus provision for auxiliary fuel tanks on the fuselage and inboard wing pylons.

Overseas air forces which include Jaguars in their inventories are those of Ecuador, Oman and India.

Airbus Industrie A300B and A310

British participation in the European wide-bodied airliner project has gone through so many vicissitudes that it is not possible to give a realistic summary of their sequence here, in the scale adopted for the other collaborative projects. Suffice it to say that the British industry's contribution to the projects, both A300B and A310, has been the design and production of their main wing structures. In the case of the former, Hawker Siddeley's Hatfield division carried out the work using company funds,

A300 Airbus at Farnborough in 1976 (Mike Bowyer).

after the withdrawal of government support in 1969. It was not until January 1979 that Britain became a full partner in Airbus Industrie, with a 20 per cent share in the company. Other countries involved, either as shareholders or associates, are France, Germany, Spain, Belgium and the Netherlands.

The first prototype of the A300B series flew on October 28 1972 and a production aircraft, in Air France colours, initiated passenger services on May 23 1974, flying from Paris to London. The 'baby brother' A310 is expected to enter service during 1983. Passenger capacities (maximum) are 336 and 255 respectively.

Order books for both airliners are in a reasonably healthy state. One of the most significant developments was the order placed in 1978 by Eastern Airlines for a fleet of 23 of the A300B4 model, plus an option for nine further aircraft. This American airline also has options for 25 of the A310. Orders from North America are particularly encouraging as the Airbus is in direct competition with the Tristar, 747 and DC-10.

Hawker Siddeley Hawk

In the world of modern aircraft development it seems usual for each new type to receive frantic coverage in the news media on the grounds of massive waste of public money, lack of need in the first place or inferiority to some other project, or a disastrous performance when at last flown after prolonged development. One aeroplane which has sailed well clear of those storms of controversy is the advanced trainer designed by Hawker Siddeley and named Hawk.

When the collaborative Jaguar metamorphosed as a full-blooded strike aircraft, the original requirement for a training aircraft had to be rewritten. As far as the British were concerned this rewrite took form as Air Staff Requirement 362, which brought forth designs from BAC and Hawker Siddeley, the latter's HS 1182 being selected in October 1971. The new type, dubbed Hawk in

The first Hawk T Mk 1 in red and white paint scheme at its 'unveiling' at Dunsfold in August 1974 (MoD).

Camouflaged Hawk flying over the Pyramids during a demonstration tour of the Middle East (Hawker Siddeley).

1972, would replace the Gnat and Hunter in the RAF, serving as a multi-talented trainer in the initial, advanced and weapons training roles.

Powered by the Jaguar's Adour engine, the first aircraft (there was no prototype) made its maiden flight from Dunsfold in the hands of Duncan Simpson on August 21 1974, and was displayed at Farnborough in September, proving to be superior in all respects to its design specification. Production was already well under way and the first aircraft were delivered to the RAF in November 1976. The Central Flying School and No 4 Flying Training School, both based at Valley, Anglesey, were the initial recipients, becoming fully effective in mid-1977. The Tactical Weapons Unit at RAF Brawdy operates the Hawk in its aggressive mode, with various weapon fits on three hardpoints; the centre-line fuselage position can take a pod-mounted 30 mm Aden gun and its ammunition, while the wing pylons may carry Matra rocket pods or practice bombs. The Hawk's total weapon capacity is 5,600 lb and for tactical support operations it would use an extra pair of wing pylons for a full load. This ability is not called upon in RAF service, but is a feature of aircraft equipped for overseas demonstration.

Finland was the first country to express interest, in 1976, with a letter of intent for a batch of 50, which involves fabrication of some components and assembly and testing of the majority of the aircraft in that country. Other countries with Hawks in service are Indonesia and Kenya, and the US Navy will be using the machine from the mid-1980s as a result of its success in the VTXTS election.

To millions the Hawk will become familiar as the mount of the 'Red Arrows' aerobatic team, whose Gnats the type replaced for the 1980 season. As the Hawk is likely to be active to the turn of the century, it may well exceed the longevity of its illustrious predecessor in that formation.

Panavia Tornado

Shortly after the end of the Second World War, Barnes Wallis, then with Vickers, Weybridge, had begun research on the variable geometry concept and from 1949 for a period of ten years, carried out exhaustive trials with a series of swing-wing models. Unfortunately this dedicated work was not followed up with full-scale designs, and it was not until after the formation of BAC, when the variable geometry research was moved to the team at Warton (formerly English Electric) in 1964, that the long road to Tornado was set foot upon.

There, a design called P.45 was initiated with a multi-role capability, covering the training, strike and interceptor roles. In 1965 the next step, in parallel with Jaguar development, was the Anglo-French Variable Geometry project (AFVG) to be produced as a joint venture by BAC and Dassault. In 1967 this, too, was abandoned after the withdrawal of French interest, but design studies were continued in Britain. By coincidence, a similar stream of work was beginning in Germany at the Messerschmitt-Bölkow-Blohm consortium, and tentative enquiries on a further collaborative project began to gain ground with the additional interest in the shared design expressed by Italy. In 1968 Belgium, the Netherlands and Canada joined the project, which by then had become MRA-75 (Multi Role Aircraft for 1975—ie, to be in service in that year).

In March 1969 the German-based company Panavia Aircraft GmbH was set up by Britain, Germany, Italy and the Netherlands (the other two partners had dropped out) and the Project Definition Phase for the Multi Role Combat Aircraft started on May 1. A Memorandum of Understanding on this was signed by three of the governments but the Netherlands withdrew from the MRCA programme in July. The shares in the company were then settled at 42½ per cent each for BAC and MBB and 15 per cent for Aeritalia. The formation of a second company to look after development of the engine was effected in September, with a head office in Bristol. This was named Turbo-Union Ltd, with shares distributed between Rolls-Royce, Motoren-und-Turbinen Union and Fiat at 40-40-20 per cent respectively.

Above right *Camouflaged prototype* 03 *flown from Warton by David Eagles and Tim Ferguson in August 1975* (BAC). **Right** *The first Tornado GR Mk 1(T) to be delivered to the Tri-national Tornado Training Establishment at Cottesmore on July 1 1980* (MAP).

Below *The red, white and black second prototype MRCA flown by Paul Millett from Warton* (BAC).

Further stages in the development of the aircraft were passed through smoothly, culminating in an agreement to start series production on July 29 1976. The name Tornado was chosen. By that date there was no doubt that the aircraft existed. Eight prototypes—three German, four British and one Italian—had been flown and the final prototype, the second Italian machine, was in the air the following February. The type's first flight was accomplished on August 14 1974 from Manching, with BAC test pilot Paul Millett in the front seat and his opposite number from MBB, Nils Meister, in the rear seat. Millett also made the maiden flight of the first British-built aircraft on October 30 1974 from Warton.

In British service there will be two versions. The GR Mk 1 is to replace the Vulcan and Buccaneer, while the specialised interceptor, F Mk 2, formerly known as the Air Defence Variant (ADV), will supersede the Lightning and Phantom. To prepare the Tornado for service, a Tri-National Tornado Training Establishment has been set up at RAF Cottesmore and the first delivery of a production aircraft, a GR Mk 1(T), to this base was made on July 1 1980.

British Aerospace Sea Harrier

As early as February 1963 the first deck landing by a V/STOL aeroplane was made when Bill Bedford took the first prototype P.1127 aboard HMS *Ark Royal* on the 8th of that month. At that time the Naval Staff were considering their requirement for a specialised maritime V/STOL aircraft which was to be an RN-oriented version of the P.1154 extrapolated from the P.1127 and with a performance reaching Mach 2.0. The successful trials aboard *Ark Royal* might have suggested that this train of development would proceed but in 1964 the naval version of P.1154 was cancelled and an order placed for a batch of McDonnell F-4K Phantoms instead.

Despite the about-turn this represented (with

Facing page *Sea Harrier FRS Mk 1s of 801 Squadron aboard HMS* Invincible (Paul Beaver and MAP).

Below *The forerunner of the Sea Harrier was the P.1127, which carried out deck landing and other trials aboard* Ark Royal *in 1963* (Paul Beaver).

N
ROYAL NAVY
005

N
ROYAL NAVY
003

regard to the RN) it did not signal the end of seagoing trials with members of the P.1127 family. In 1967 a Harrier made landings on the helicopter platform of the Italian cruiser *Andrea Doria* and this performance was repeated from the similarly configured *Blake* of the Royal Navy in 1969. These were but two examples of Harrier trials at sea; another, from the Spanish aircraft carrier *Dédalo*, was instrumental in the adoption of the type, as the Matador, by the Spanish Navy.

It was not until 1970 that full-scale trials with Harriers from the conventional carrier *Eagle* were made and by then the days of the fixed-wing element of the Fleet Air Arm were numbered. The hoped-for advanced carrier, known as CVA-01, had been cancelled in 1966 and the existing ships would be phased out by the end of the 1970s. Reliance for air support of the fleet would fall upon the RAF and the FAA role would be reduced to providing an anti-submarine guard of helicopters. For this purpose a new class of ship, known variously as the 'through-deck cruiser', 'anti-submarine cruiser' or 'command cruiser', would be ordered to operate as the flagship of an anti-submarine group equipped with Sea King a/s helicopters.

However, it soon became apparent that the RAF, itself suffering from severe economic restraints, would be unable to provide an effective land-based cover for naval operations. Therefore, in 1972, design work on a Maritime Harrier was started to provide an aircraft capable of operating from the new command cruiser. Further budgetary difficulties, industrial and energy crises and two general elections in 1974 conspired to push back the date for a firm decision on development go-ahead. In May 1975 the announcement was made in the House of Commons by Defence Minister Roy Mason, that a batch of 24 Sea Harriers and two further 'anti-submarine cruisers' would be built.

The aircraft would be a multi-role development of the standard land-based Harrier with adaptations to suit it to the salt environment (ie, replacing of corrosion-prone materials) and with new equipment to fit it for the different tasks it would perform. The prime role would be that of fighter, forming a defensive umbrella for the anti-submarine group's activities, with secondary capacity for reconnaissance (of enemy naval units or formations) and strike, hence the designation FRS Mk 1.

The major change of the Sea Harrier's airframe concerns the fuselage forward of the air intakes. The nose contains the Ferranti 'Blue Fox' radar for use in the air-to-air or air-to-surface modes, the radome being hinged for folding to port to assist in striking down on board ship. The cockpit is raised 11 in to provide a better all-round view to the pilot and its layout is revised to allow the best arrangement of the instruments and incorporation of the radar display. Raising the cockpit also allows extra avionics equipment to occupy the extra space below it. Armament fits include the standard Harrier pod-mounted 30 mm Aden cannon plus Sidewinder AAMs, Martel air-to-surface missiles, bombs or rockets on the four wing and one fuselage centre-line (bombs only) pylons. Drop tanks may also be carried.

The first aircraft flew from Dunsfold on August 20 1978 in the hands of John Farley and by November that year sea trials had commenced aboard HMS *Hermes*, involving five Sea Harriers. Two of these were from the Yeovilton-based No 700A Intensive Flying Trials Unit, which has since become No 899 Squadron, the type's shore-based parent unit.

On March 19 1980 HMS *Invincible*, the first of the new carriers, was handed over to the Royal Navy at Portsmouth, commissioning the following July, and the first front-line squadron, No 800, was commissioned at Yeovilton on St George's Day (April 23). The culmination of this renaissance of naval fixed-wing flying came on May 20 1980 when a Sea Harrier landed for the first time on HMS *Invincible* at sea, and its vindication in 1982 during the Falklands confrontation.

British Aerospace BAe 146

The BAe 146 is a small short-range airliner with a 3,500-ft take-off run, allowing for operation into and out of smaller airfields. As these airfields may have unpaved runways the wing is mounted high on the fuselage to keep the engines clear of debris thrown up during take-off.

Design studies in Britain for a feederliner began in the late 1950s with the DH 123 by de Havilland and continued over the ensuing decade, through the aero-industry amalgamations, until the HS 144 design of 1969 which brought together the best features of the current de Havilland and Avro projects. As it was important to achieve low noise levels, the engines would have to be turbofans, in the case of the HS 144 a pair of the then-forthcoming Rolls-Royce RB.203 Trents mounted on the sides of the fuselage behind the wing.

With the collapse of the Rolls-Royce company in 1970, Hawker Siddeley were left without engines, which delayed advancement of the design while a new powerplant was tracked down. In 1971, after selecting the American Avco-Lycoming ALF502, a revised design called HS 146 was drawn up featuring four of these engines underslung from a high wing. Further refinement took place over the next 12 months and Hawker Siddeley sought government support for the project. This was forthcoming in the form of a 50 per cent sharing of the projected development costs of £120m and the official announcement of the project was made on August 29 1973.

Unfortunately for all concerned the fuel crisis of 1974 led to the shelving of the programme, owing to the uncertainty of the economic situation that this caused. It was not until July 1978, after the aircraft industry had become the national concern of British Aerospace, that the government announcement was made of the continuation of the project.

At present two variants are under development, the Series 100 with a capacity for 71-88 passengers in five- or six-abreast seating and the slightly stretched Series 200 with seating for 80-110. There is also a potential for military use as a medium transport aircraft with the ability to carry a light tank or a 1-ton truck or light field gun. A design, based on the Series 200, with rear fuselage loading doors has been drawn up by the Manchester division of BAe, under the designation of 146M.

The first aircraft, *G-SSSH*, made a successful maiden flight on September 3 1981, and healthy interest in Britain's first indigenous commercial type since the 1-11 of 1963 has been expressed by several operators.

One of the first BAe 146s which received '. . . you know who . . .' registrations (Darby Childerhouse, BAe).

Appendices

1 Aircraft data

Type	Span (ft in)	Length (ft in)	Height (ft in)	Wing area (sq ft)	Engines	Thrust (lb each engine)	All-up weight (lb)	Max speed (mph) (at altitude)	Service ceiling (ft)	Range (miles)
1. Fighters; including trainer and recce variants and prototypes										
Gloster E.28/39	29 0	25 $3\frac{3}{4}$	9 3	146.5	1 × Power Jets W.1	860	3,700	366 at 15,000	32,000	460
Gloster Meteor I	43 0	41 3	13 0	374	2 × RR Welland I	1,700	13,800	410 at 10,000	40,000	500
Gloster Meteor III	43 0	41 3	13 0	374	2 × RR Welland I	2,000	13,920	415 at 10,000	40,000	500
Gloster Meteor F Mk 4	37 2	41 0	13 0	350	2 × RR Derwent V	3,500	15,175	585 at sea level	50,000	600
Gloster Meteor T Mk 7	37 2	43 6	13 0	350	2 × RR Derwent 5 or 8	3,500	17,600	590 at 10,000	45,000	750 (extra fuel)
Gloster Meteor F Mk 8	37 2	44 7	13 0	350	2 × RR Derwent 8	3,500	19,100	590 at sea level	44,000	980
Gloster Meteor FR Mk 9	37 2	44 7	13 0	350	2 × RR Derwent 8	3,500	15,660	595 at 10,000	44,000	690
Gloster Meteor PR Mk 10	43 0	44 3	13 0	374	2 × RR Derwent 8	3,500	15,330	575 at 10,000	47,000	1,400
Armstrong Whitworth Meteor NF Mk 11	43 0	48 6	13 11	374	2 × RR Derwent 8	3,500	20,000	554 at 10,000	40,000	920
Armstrong Whitworth Meteor NF Mk 14	43 0	51 4	13 11	374	2 × RR Derwent 9	3,800	21,200	578 at 10,000	40,000	920
de Havilland Vampire I	40 0	30 9	8 10	266	1 × DH Goblin 2	3,100	10,480	540 at 20,000	?	730
de Havilland Vampire F Mk 3	40 0	30 9	8 10	266	1 × DH Goblin 2	3,100	11,970	531 at 20,000	43,500	1,145
de Havilland Vampire FB Mk 5	38 0	30 9	8 10	262	1 × DH Goblin 2	3,100	12,360	535 at 20,000	40,000	1,170
de Havilland Vampire FB Mk 9	38 0	30 9	8 10	262	1 × DH Goblin 3	3,350	12,390	548 at 30,000	42,800	1,220
de Havilland Vampire NF Mk 10	38 0	34 7	6 7	261	1 × DH Goblin 3	3,350	11,350	550 at 20,000	40,000	1,220
de Havilland Vampire T Mk 11	38 0	34 $6\frac{1}{2}$	6 2	262	1 × DH Goblin 35	3,500	11,150	549 at 20,000	40,000	840
de Havilland Sea Vampire F Mk 20	38 0	30 9	8 10	262	1 × DH Goblin 2	3,100	12,660	526 at 20,000	43,500	1,145
de Havilland Sea Vampire T Mk 22	38 0	34 $6\frac{1}{2}$	6 2	262	1 × DH Goblin 35	3,500	11,150	538 at sea level	40,000	840
de Havilland Venom FB Mk 1	41 8	31 10	6 2	279.75	1 × DH Ghost 103	4,850	15,400	640	48,000	1,075
de Havilland Venom NF Mk 3	42 11	36 8	6 6	279.75	1 × DH Ghost 104	4,950	14,620	630	49,200	1,000
de Havilland Sea Venom FAW Mk 22	42 11	36 7	8 $6\frac{1}{4}$	279.75	1 × DH Ghost 105	5,300	15,800	575 at sea level	40,000	705
Gloster E.1/44	36 0	38 0	11 8	254	1 × RR Nene 2	5,000	11,470	620 at sea level	48,000	?
Saunders-Roe SR.A/1	46 0	50 0	16 9	415	2 × MetroVick F.2/4 Beryl	3,850	16,255	512 at sea level	?	482
Supermarine Attacker F Mk 1	36 11	37 6	9 11	226	1 × RR Nene 3	5,100	11,500	590 at sea level	45,000	1,190
Hawker P.1040	36 6	39 8	?	?	1 × RR Nene 1	4,500	?	?	?	?
Hawker P.1072	36 6	37 7	?	?	1 × RR Nene 2 1 × AS Snarler 1	5,000 2,000	14,050	M 0.82	47,000	?
Hawker P.1052	31 6	39 7	10 6	258	1 × RR Nene 2	5,000	13,488	650	45,500	?

Type	Span (ft in)	Length (ft in)	Height (ft in)	Wing area (sq ft)	Engines	Thrust (lb each engine)	All-up weight (lb)	Max speed (mph) (at altitude)	Service ceiling (ft)	Range (miles)
Hawker P.1081	31 6	37 4	10 10	258	1 × RR Nene 2	5,000	14,480	M 0.89 at 36,000 695 at sea level	47,000	?
Hawker Sea Hawk FGA Mk 6	39 0	39 8	8 9	278	1 × RR Nene 103	5,200	16,200	524 at 10,000	44,500	288
Supermarine 510	31 8½	38 1	8 9¾	273	1 × RR Nene 2	5,000	12,177	650	?	?
Supermarine 535	32 4	?	?	?	1 × RR Nene 2	5,000	?	?	?	?
Supermarine 541	32 4	?	?	?	1 × RR Avon	7,000	?	?	?	?
Supermarine Swift F Mk 1	32 4	41 5½	12 6	306	1 × RR Avon 13	7,500	?	?	?	?
Supermarine Swift FR Mk 5	32 4	42 3	13 6	?	1 × RR Avon 114	7,175 9,450 (i)	21,400	685 at sea level	?	480
Hawker Hunter F Mk 1	33 8	45 10½	?	?	1 × RR Avon 13	7,550	16,200	M 0.93 at 36,000	48,800	?
Hawker Hunter F Mk 2	33 8	45 10½	?	?	1 × AS Sapphire 101	8,000	16,200	M 0.94 at 36,000	50,000	?
Hawker Hunter F Mk 6	33 8	45 10½	13 2	340	1 × RR Avon 203	10,050	17,750	M 0.95 at 36,000 715 at sea level	51,500	1,840
Hawker Hunter T Mk 7	33 8	48 10½	13 2	349	1 × RR Avon 122	7,575	17,200	615 at 40,000 (M 0.93)	47,000	?
Hawker Hunter FGA Mk 9	33 8	45 10½	13 2	349	1 × RR Avon 207	10,150	18,000	M 0.95 at 36,000 715 at sea level	51,500	300 (tactical radius)
Hawker Hunter FR Mk 10	33 8	46 1	13 2	349	1 × RR Avon 207	10,150	18,090	M 0.95 at 36,000	51,500	1,840
Hawker Hunter GA Mk 11	33 8	45 10½	13 2	?	1 × RR Avon 113	7,500	17,100	M 0.94 at 36,000	50,000	?
Supermarine 508	41 0	50 0	11 7½	310	2 × RR Avon 3	6,500	?	?	?	?
Supermarine 529	41 0	50 0	11 7½	310	2 × RR Avon 3	6,500	22,584	M 0.92	?	?
Supermarine 525	38 6	55 0	?	?	2 × RR Avon 114	7,175	28,169	?	?	?
Supermarine Scimitar F Mk 1	37 2	55 3	17 4	484.9	2 × RR Avon 202	11,250	40,000	710 at 10,000	46,000	650
de Havilland DH 110	51 0	52 1½	11 0	648	2 × RR Avon 13	7,550	32,000	M 0.9	45,000	?
de Havilland Sea Vixen FAW Mk 1	51 0	55 7	10 9	648	2 × RR Avon 208	11,230	35,000	645 at 10,000	48,000	?
de Havilland Sea Vixen FAW Mk 2	51 0	55 7	10 9	648	2 × RR Avon 208	11,230	37,000	640 at 10,000	48,000	?
Gloster Javelin FAW Mk 1	52 0	56 3	16 0	927	2 × AS Sapphire 6	8,000	36,690	M 0.94 at 40,000	52,500	?
Gloster Javelin T Mk 3	52 0	59 11	16 0	927	2 × AS Sapphire 6	8,000	42,000	M 0.91 at 35,000	46,000	?
Gloster Javelin FAW Mk 7	52 0	56 3	16 0	927	2 × AS Sapphire 7	11,000	40,270	M 0.94 at 45,000	52,800	960
Gloster Javelin FAW Mk 9	52 0	56 9	16 0	927	2 × AS Sapphire 7R	11,000 12,300 (i)	43,165	M 0.92 at 35,000	52,000	930
English Electric (BAC) Lightning F Mk 1A	34 10	55 3	19 7	380	2 × RR Avon 210	11,250 14,430 (i)	40,000	1,500 at 36,000 (M 2.3)	60,000	1,500

Type	Span (ft in)	Length (ft in)	Height (ft in)	Wing area (sq ft)	Engines	Thrust (lb each engine)	All-up weight (lb)	Max speed (mph) (at altitude)	Service ceiling (ft)	Range (miles)
English Electric (BAC) Lightning F Mk 6	34 10	55 3	19 7	460	2 × RR Avon 301	12,690 16,360 (i)	45,000	1,500 at 40,000	60,000	800 (vi)
Folland Midge	20 8	27 9	8 9	125	1 × AS Viper 101	1,640	4,900	550 at sea level	40,000	?
Saunders-Roe SR 53	28 1 (ii) 25 $1\frac{1}{4}$	45 10	10 10	274	1 × AS Viper 1 × DH Spectre	1,750 8,000	12,780	M 2.2	?	?
Hawker Siddeley Kestrel F (GA)Mk 1	22 11	42 6	10 9	186	1 × BS Pegasus 5	15,200	15,000	690 at sea level	50,000	350
Hawker Siddeley Harrier GR Mk 1	29 8 (iii) 25 3	45 6	10 9	216 201	1 × RR Pegasus 101	19,000	19,500	730	50,000	320
Hawker Siddeley Harrier T Mk 2	25 3	55 $9\frac{1}{2}$	13 8	201	1 × RR Pegasus 101	19,000	25,000	730	50,000	390
Hawker Siddeley Harrier GR Mk 3	25 3	46 4	10 9	201	1 × RR Pegasus 103	21,500	25,000	737 at low alt (M 0.95)	50,000	400
British Aerospace (HS) Sea Harrier FRS Mk 1	25 3	47 7	12 2	201	1 × RR Pegasus 104	21,500	25,000	735 at low alt	50,000	100 (tactical radius)
Sepecat Jaguar GR Mk 1	28 6	50 11	16 $0\frac{1}{2}$	260	2 × RR Turbomeca Adour 102	4,620 7,140 (i)	32,600	1,057 at 32,810 (M 1.6)	?	710
2. Trainers										
Hunting Jet Provost T Mk 1	35 2	31 11	12 8	?	1 × AS Viper 5	1,640	6,750	330 at 20,000	?	?
Hunting Jet Provost T Mk 3	36 11 (iv)	32 5	10 2	?	1 × BS Viper 8	1,750	7,200	326 at 30,000	33,000	565
Hunting Jet Provost T Mk 5	35 4 (v)	33 $7\frac{1}{2}$	10 2	213.7	1 × BS Viper 11	2,500	9,200	440 at 25,000	36,700	900
BAC Strikemaster	36 11	33 $8\frac{1}{2}$	10 2	213.7	1 × RR Viper 20	3,410	9,200 (vii)	472 at 20,000	40,000	640
Miles Student	29 2	31 6	6 3	144	1 × Blackburn Turboméca Marboré 2	883	3,600	298 at 20,000	?	621
Folland (HS) Gnat T Mk 1	24 0	31 9	9 $7\frac{1}{2}$	175	1 × BS Orpheus 100	4,230	8,630	M 0.97	50,000	1,180
Hawker Siddeley (BAe) Hawk T Mk 1	30 $9\frac{3}{4}$	38 11	13 5	179.6	1 × RR Turboméca Adour 151	5,340	16,500	576 at 30,000	45,000	1,300 (ferry)
3. Experimental types										
de Havilland DH 108 (third aircraft)	39 0	26 $9\frac{1}{2}$	?	328	1 × DH Goblin 4	3,750	?	M1.0 in a dive	?	?
Short SB 5 (50° sweep back)	35 $2\frac{1}{4}$	45 9	16 7	273	1 × RR Derwent	3,500	12,000	311	?	?
Short SB 5 (69° sweep back)	25 $11\frac{3}{4}$	45 9	?	282	1 × BS Orpheus	?	?	?	?	?
Short Sherpa	38 0	31 10	9 1	230	2 × Blackburn Turboméca Palas	350	3,268	170	?	?
Handley Page HP 88	40 0	40 0	12 6	?	1 × RR Nene 3	5,100	14,460	M 0.85	?	?
Avro 707A	34 2	42 4	11 7	?	1 × RR Derwent 8	3,600	9,800	?	?	?
Boulton Paul P.111	25 0 33 $5\frac{1}{2}$ (viii)	26 1	12 $6\frac{1}{2}$	200	1 × RR Nene 3	5,100	11,400	M 0.95	?	?
Fairey FD 1	19 $6\frac{1}{2}$	26 3	?	?	1 × RR Derwent 8	3,600	6,800	345	?	?
Fairey FD 2	26 10	51 $7\frac{1}{2}$	11 0	360	1 × RR Avon 5	12,000 (i)	13,400	1,147 at 38,000	?	?
BAC 221	25 0	57 $7\frac{1}{2}$	13 0	504	1 × RR Avon 28R	14,000 (i)	?	1,060	?	?

Type	Span (ft in)	Length (ft in)	Height (ft in)	Wing area (sq ft)	Engines	Thrust (lb each engine)	All-up weight (lb)	Max speed (mph) (at altitude)	Service ceiling (ft)	Range (miles)
Handley Page HP 115	20 0	45 0	?	432	1 × BS Viper 9	1,900	5,050	248	?	?
Bristol 188	35 1	71 0 (ix)	13 4	396	2 × BS Gyron Junior 10R	14,000 (i)	?	1,200	?	?
Hunting H.126	45 4	50 2 (x)	15 6	?	1 × BS Orpheus	?	?	?	?	?
Short SC 1	23 6	29 10	10 8	141.9	5 × RR RB.108	2,130	8,050	246 at sea level	?	?
Hawker P.1127	24 4	49 0	?	?	1 × BS Pegasus 3	13,500	11,800 (VTO)	M 0.87	?	?
Armstrong Whitworth AW 52 (first prototype)	90 0	37 4	14 5	1,314	2 × RR Nene 2	5,000	31,500	488	50,000	780
Vickers Nene Viking	89 3	65 2	19 6	882	2 × RR Nene 1	5,000	33,500	468 at 10,000	44,000	345
Vickers Tay Viscount	89 0	74 6	26 3	885	2 × RR Tay 1	?	40,000	?	?	?
Avro Ashton	120 0	89 6½	31 3	1,421	4 × RR Nene 5/6	5,000	72,000	438 at 20,000	40,500	1,725
4. Bombers, including recce variants and prototypes										
English Electric (BAC) Canberra B Mk 2	63 11½	65 6	15 7	960	2 × RR Avon 101	6,500	46,000	570 at 40,000	48,000	2,660
English Electric (BAC) Canberra PR Mk 7	63 11½	66 8	15 7	960	2 × RR Avon 109	7,500	55,000	580 at 40,000	48,000	4,340
English Electric (BAC) Canberra B (I)Mk 8	63 11½	65 6	15 7	960	2 × RR Avon 109	7,500	50,992	541 at 40,000	48,000	805
English Electric (BAC) Canberra PR Mk 9	67 10	?	?	?	2 × RR Avon 206	10,050	?	?	60,000	?
Short Sperrin	109 0	102 3	28 6	1,897	4 × RR Avon 13	7,500	115,000	564 at 15,000	42,000	3,860
Vickers Valiant B Mk 1	114 4	108 3	32 2	2,362	4 × RR Avon 201	10,000	140,000	567 at 30,000	54,000	4,500
Avro (HS) Vulcan B Mk 1	99 0	97 1	26 1	3,554	4 × BS Olympus 101	11,000	90,000	580 at 48,000	60,000	4,500
Avro (HS) Vulcan B Mk 2	111 0	99 11	27 2	3,964	4 × BS Olympus 201	17,000	225,000	620 at 55,000	65,000	4,600
Handley Page Victor B Mk 1	110 0	114 11	28 1½	2,406	4 × AS Sapphire 7	11,000	205,000	640 at 40,000 (M 0.92)	56,000	4,000
Handley Page Victor B Mk 2	120 0	114 11	30 1½	2,597	4 × RR Conway 201	19,750	216,000	640 at 40,000	60,000	3,000
Blackburn (HS) Buccaneer S Mk 1	44 0	63 5	16 3	515	2 × DH Gyron Junior 101	7,100	45,000	720 at sea level (M 0.95)	?	500 (tactical radius)
Blackburn (HS) Buccaneer S Mk 2	44 0	63 5	16 3	515	2 × RR Spey 101	11,100	50,000	645 at 200 (M 0.85)	?	600 (tactical radius)
BAC TSR 2	37 1½	86 0	23 5	700	2 × BS Olympus 22R	30,610 (i)	95,900 (xi)	M 2.25 (xi)	60,000	1,727 (xi)
Panavia Tornado GR Mk 1	45 7¼ 28 2½ (xii)	54 9½	18 8½	332.9	2 × Turbo Union RB.199-34R Mk 101	15,000	60,000	1,320 at 36,000 (M 2.2)	50,000	?
Hawker Siddeley Nimrod MR Mk 1	114 10	126 9	29 8½	2,121	4 × RR Spey 250	12,160	178,000	575	?	5,180

Type	Span (ft in)	Length (ft in)	Height (ft in)	Wing area (sq ft)	Engines	Thrust (lb each engine)	All-up weight (lb)	Max speed (mph) (at altitude)	Service ceiling (ft)	Range (miles)
5. Civil aircraft, including military variants										
de Havilland Comet 1	115 0	93 0	28 4	2,015	4 × DH Ghost 50	4,450	107,000	490 at 35,000	45,000	1,750
de Havilland Comet C Mk 2	115 0	96 0	28 4	2,015	4 × RR Avon 117	7,300	120,000	490 at 40,000	50,000	2,100
de Havilland Comet 4B	107 10	118 0	28 6	2,059	4 × RR Avon 524	10,500	157,960	532 at 23,500	50,000	2,500
Hawker Siddeley Trident 1	89 10	114 9	27 0	1,358	3 × RR Spey 505	9,850	117,300	589	?	930
Hawker Siddeley Trident 3	98 0	131 2	28 3	1,493	3 × RR Spey 512 1 × RR RB.162-86 booster	11,930 5,250	150,000	581	?	1,150
Vickers (BAC) VC 10	146 2	158 8	39 6	2,851	4 × RR Conway 540	20,370	314,000	568	?	5,040
Vickers (BAC) VC10 C Mk 1	146 2	133 8 (ix)	40 0	2,932	4 × RR Conway 301	22,500	323,000	580 at 30,000	?	3,668 (xiii)
Vickers (BAC) Super VC 10	146 2	171 8	39 6	2,932	4 × RR Conway 550	21,800	335,000	581	?	4,720
BAC One-Eleven Srs 200	88 6	93 6	24 6	1,003	2 × RR Spey 506	10,330	79,000	540 at 21,000	?	875
BAC One-Eleven Srs 500	93 6	107 0	24 6	1,031	2 × RR Spey 512	12,550	104,500	541 at 21,000	?	1,705
BAC One-Eleven Srs 670	93 6	93 6	24 6	1,031	2 × RR Spey 512	12,550	92,000	550 at 21,000	?	1,250
BAC/Aérospatiale Concorde	83 10	203 9	37 1	3,856	4 × RR Snecma Olympus 593	38,050	407,994	1,354 at 51,300 (M 2.04)	60,000	3,870
Airbus Industrie A300B4	147 1	175 11	54 2	2,800	2 × General Electric CF6-50C	51,000	347,230	570 at 30,000	?	2,994
Airbus Industrie A310	144 0	153 1	51 10	2,356	2 × General Electric CF6-80A	48,000	291,000 (xi)	515 at 35,000 (xi)	?	3,470 (xi)
British Aerospace BAe 146 Srs 100	86 6	85 10	28 3	832	4 × Avco-Lycoming ALF502R-3	6,700	73,850 (xi)	473 at 22,000 (xi)	?	1,385 (xi)
Hawker Siddeley (BAe) HS 125 Dominie T Mk 1	47 0	47 5	16 6	353	2 × BS Viper 520	3,310	20,500	472 at 25,000	40,000	1,338
Hawker Siddeley (BAe) HS 125 Srs 600	47 0	50 5¾	17 3	353	2 × RR Viper 601	3,750	25,000	503 at 40,000	?	1,700
Hawker Siddeley (BAe) HS 125 Srs 700	47 0	50 8½	17 7	353	2 × Garrett TFE371	3,700	24,200	494	41,000	2,400
6. Racing aircraft										
Miles Sparrowjet	28 6	30 9½	?	156	2 × Turboméca Palas	330	2,400	230	?	270

Notes

(i) Reheat; (ii) With missiles; (iii) Ferry tips; (iv) Over tip tanks; (v) Without tip tanks; (vi) With ventral tank; (vii) Less weapons; (viii) Extended tips; (ix) Excluding nose probe; (x) Including probe; (xi) Estimated; (xii) Fully swept; (xiii) With heavy payload.

Key to engine manufacturers

RR = Rolls-Royce; DH = de Havilland (taken over by BS in 1961); AS = Armstrong Siddeley (amalgamated with Bristol Aero Engines in 1958 and new company named Bristol Siddeley); BS = Bristol Siddeley (bought out by RR in 1966).

2 A summary of the military service of British jet aircraft

NB This list is not intended to be a definitive list of every unit ever to use jet aircraft in service but a guide to the major unit allocations; in the case of the more prolific Royal Air Force types, after the squadron number is given a list of the marks used in quantity by that unit.

Abbreviations

ANS	Air Navigation School
CAW	College of Air Warfare
CFS	Central Flying School
ECM	Electronic Countermeasures
ELINT	Electronic Intelligence
FRU	Fleet Requirements Unit
FTS	Flying Training School
LTF	Lightning Training Flight
OCU	Operational Conversion Unit
RAFFC	Royal Air Force Flying College
TES	Tri-partite Evaluation Squadron
TTTE	Tri-national Tornado Training Establishment
TWU	Tactical Weapons Unit

1. Gloster Meteor

1 Sqn	3, 4, 8
19 Sqn	4, 8
34 Sqn	8
41 Sqn	4, 8
43 Sqn	4, 8
45 Sqn	
54 Sqn	8
56 Sqn	3, 4, 8
63 Sqn	3, 4, 8
64 Sqn	8
65 Sqn	8
66 Sqn	3, 4, 8
72 Sqn	8
74 Sqn	3, 4, 8
91 Sqn	3
92 Sqn	3, 4, 8
111 Sqn	8
124 Sqn	3
222 Sqn	3, 4, 8
234 Sqn	3
245 Sqn	3, 4, 8
247 Sqn	8
257 Sqn	3, 4, 8
263 Sqn	3, 4, 8
266 Sqn	3, 4
500 Sqn	3, 4, 8
501 Sqn	8
504 Sqn	3, 4, 8
600 Sqn	4, 8
601 Sqn	8
604 Sqn	8
609 Sqn	8
610 Sqn	4, 8
611 Sqn	4, 8
615 Sqn	4, 8
616 Sqn	I, 3, 4, 8

Reconnaissance Meteors

2 Sqn
13 Sqn
79 Sqn
81 Sqn
208 Sqn
541 Sqn

2. Armstrong Whitworth Meteor

5 Sqn	11
11 Sqn	11
25 Sqn	12, 14
29 Sqn	11
33 Sqn	14
39 Sqn	13
46 Sqn	12, 14
60 Sqn	14
64 Sqn	12, 14
68 Sqn	11
72 Sqn	12, 14
85 Sqn	11, 12, 14
87 Sqn	11
96 Sqn	11
125 Sqn	11
141 Sqn	11
151 Sqn	11
152 Sqn	12, 14
153 Sqn	12, 14
219 Sqn	13
256 Sqn	11
264 Sqn	11, 14

3. de Havilland Vampire

3 Sqn	1, 5
4 Sqn	5
5 Sqn	5
6 Sqn	5
8 Sqn	9
11 Sqn	5
14 Sqn	5
16 Sqn	3, 5
20 Sqn	9
23 Sqn	10
25 Sqn	10
26 Sqn	5, 9
28 Sqn	5, 9
32 Sqn	3, 5, 9
45 Sqn	9
54 Sqn	1, 3, 5
60 Sqn	5, 9
67 Sqn	5
71 Sqn	5
72 Sqn	1, 3, 5
73 Sqn	3, 5, 9
93 Sqn	5
94 Sqn	5
98 Sqn	5
112 Sqn	5
118 Sqn	5
130 Sqn	1, 5
145 Sqn	5
151 Sqn	10
185 Sqn	5
213 Sqn	5, 9
234 Sqn	5, 9
247 Sqn	1, 3, 5
249 Sqn	5, 9
266 Sqn	5
501 Sqn	1, 5
502 Sqn	1, 5, 9
601 Sqn	3
602 Sqn	5
603 Sqn	5
604 Sqn	3
605 Sqn	1, 5
607 Sqn	5
608 Sqn	3, 5, 9
612 Sqn	5
613 Sqn	5
614 Sqn	3, 5, 9

Vampire trainers

1 FTS
4 FTS
5 FTS
7 FTS
8 FTS

4. de Havilland Venom

5 Sqn	1, 4
6 Sqn	1, 4
8 Sqn	1, 4
11 Sqn	1, 4
14 Sqn	1
16 Sqn	1
23 Sqn	2, 3
28 Sqn	1, 4
32 Sqn	1
33 Sqn	2
45 Sqn	1
60 Sqn	1, 4
73 Sqn	1
89 Sqn	3
94 Sqn	1
98 Sqn	1
118 Sqn	1
125 Sqn	3
141 Sqn	3
145 Sqn	1
151 Sqn	3
208 Sqn	4
219 Sqn	2
249 Sqn	1, 4
253 Sqn	2
266 Sqn	1, 4

5. Hawker Hunter

1 Sqn	5, 6, 9
2 Sqn	10
3 Sqn	4
4 Sqn	4, 10
8 Sqn	9
14 Sqn	4, 6
19 Sqn	6
20 Sqn	4, 6, 9
26 Sqn	4, 6
28 Sqn	9
34 Sqn	5
41 Sqn	5
43 Sqn	1, 4, 6, 9
45 Sqn	9
54 Sqn	1, 4, 6, 9
56 Sqn	5, 6
58 Sqn	9
63 Sqn	6, 9
65 Sqn	6
66 Sqn	4, 6, 9
67 Sqn	4
71 Sqn	4
74 Sqn	4, 6
79 Sqn	9
92 Sqn	4, 6
93 Sqn	4, 6
98 Sqn	4
111 Sqn	4, 6
112 Sqn	4
118 Sqn	4, 6
130 Sqn	4
208 Sqn	6, 9
222 Sqn	1, 4
234 Sqn	4, 9
245 Sqn	4
247 Sqn	1, 4, 6
257 Sqn	2, 5
263 Sqn	2, 5, 6
229 OCU	
237 OCU	
4 FTS	
1 TWU	
2 TWU	

6. Gloster Javelin
3 Sqn 4
5 Sqn 5, 9
11 Sqn 9
19 Sqn 2, 2A
23 Sqn 4, 7, 9
25 Sqn 7, 9
29 Sqn 6, 9
33 Sqn 7
41 Sqn 4, 8
46 Sqn 1, 2
60 Sqn 9
64 Sqn 7, 9
72 Sqn 4
85 Sqn 2, 6, 8
87 Sqn 1
89 Sqn 2, 6
141 Sqn 4
151 Sqn 5
228 OCU

7. Supermarine Swift
2 Sqn 5
56 Sqn 1, 2
79 Sqn 5

8. English Electric (BAC) Lightning
5 Sqn 3, 6
11 Sqn 3, 6
19 Sqn 2, 2A
23 Sqn 3, 6
29 Sqn 3
56 Sqn 1A, 3, 6
74 Sqn 1, 3, 6
92 Sqn 2, 2A
111 Sqn 1A, 3, 6
226 OCU
LTF

9. Hawker Siddeley Kestrel
TES

10. Hawker Siddeley Harrier
1 Sqn
3 Sqn
4 Sqn
20 Sqn
233 OCU

11. Sepecat Jaguar
2 Sqn
6 Sqn
14 Sqn
17 Sqn
20 Sqn
31 Sqn
41 Sqn
54 Sqn
226 OCU

12. English Electric (BAC) Canberra
3 Sqn 8
6 Sqn 2, 6, 16
9 Sqn 2, 6
10 Sqn 2
12 Sqn 2, 6
14 Sqn 8
XV Sqn
16 Sqn 8
18 Sqn 2
21 Sqn 2, 6
27 Sqn 2
32 Sqn 2, 15
35 Sqn 2
40 Sqn 2
44 Sqn 2
45 Sqn 2, 15
50 Sqn 2
57 Sqn 2
59 Sqn 2, 8
61 Sqn 2
73 Sqn 2, 15
76 Sqn 6
88 Sqn 8
90 Sqn 2
100 Sqn 2, 6, 8
101 Sqn 2, 6
102 Sqn 2
103 Sqn 2
104 Sqn 2
109 Sqn 2, 6
115 Sqn 2
139 Sqn 2, 6
149 Sqn 2
199 Sqn 2
207 Sqn 2
213 Sqn 6
249 Sqn 2, 6, 16
617 Sqn 2, 6
Reconnaissance Canberras
7 Sqn (TT)
13 Sqn
17 Sqn
31 Sqn
39 Sqn
58 Sqn
80 Sqn
81 Sqn
82 Sqn
100 Sqn (TF)
540 Sqn
542 Sqn
Other units
360 Sqn ECM Training
51 Sqn ELINT
231 OCU

13. Vickers Valiant
7 Sqn
18 Sqn
49 Sqn
90 Sqn
138 Sqn
148 Sqn
199 Sqn
207 Sqn
214 Sqn
543 Sqn Recce
232 OCU

14. Avro (HS) Vulcan
9 Sqn 2
12 Sqn 2
27 Sqn 2
35 Sqn 2
44 Sqn 1, 1A, 2
50 Sqn 1, 1A, 2
83 Sqn 1, 2
101 Sqn 1, 1A, 2
617 Sqn 1, 2
27 Sqn Recce
230 OCU

15. Handley Page Victor
10 Sqn 1
15 Sqn 1, 1A
55 Sqn 1A
57 Sqn 1, 1A
100 Sqn 1A, 2
139 Sqn 2
543 Sqn Recce
55 Sqn K1A, K2
57 Sqn K1, K2
214 Sqn K1
232 OCU

16. Hawker Siddeley Buccaneer
12 Sqn
15 Sqn
16 Sqn
208 Sqn
216 Sqn
237 OCU

17. Panavia Tornado
TTTE
9 Sqn

18. Hawker Siddeley Nimrod
42 Sqn
120 Sqn
201 Sqn
203 Sqn
206 Sqn
51 Sqn ELINT
236 OCU

19. Hunting (BAC) Jet Provost
1 FTS
2 FTS
3 FTS
6 FTS
7 FTS
CFS
1 TWU
CAW
Cranwell

20. Folland (HS) Gnat
4 FTS
CFS

21. Hawker Siddeley (BAe) Hawk
4 FTS
CFS
1 TWU
2 TWU
63 Sqn
151 Sqn
234 Sqn

22. Hawker Siddeley (BAe) Dominie/HS 125
1 ANS
6 FTS
Cranwell
32 Sqn Communications
RAFFC

23. de Havilland (HS) Comet
216 Sqn
51 Sqn ELINT
192 Sqn ELINT

24. Vickers (BAC) VC 10
10 Sqn
241 OCU

NAVAL SQUADRONS

25. de Havilland Sea Vampire
700 Sqn
702 Sqn
736 Sqn
738 Sqn
764 Sqn
781 Sqn
787 Sqn

26. de Havilland Sea Venom
808 Sqn
809 Sqn
831 Sqn
890 Sqn
891 Sqn
892 Sqn
893 Sqn
894 Sqn

895 Sqn
736 Sqn
738 Sqn
750 Sqn
766 Sqn

27. Supermarine Attacker
800 Sqn
803 Sqn
890 Sqn
702 Sqn
736 Sqn
767 Sqn
771 Sqn
787 Sqn
1831 Sqn
1832 Sqn
1833 Sqn
1834 Sqn
1835 Sqn

28. Hawker Sea Hawk
800 Sqn
801 Sqn
802 Sqn
803 Sqn
804 Sqn
806 Sqn
807 Sqn
810 Sqn
811 Sqn
895 Sqn
897 Sqn
898 Sqn
899 Sqn
700 Sqn
736 Sqn
738 Sqn
764 Sqn
767 Sqn
1832 Sqn
1835 Sqn
1836 Sqn

29. Hawker Hunter
738 Sqn
759 Sqn
764 Sqn

30. Supermarine Scimitar
700X Sqn
800 Sqn
803 Sqn
804 Sqn
807 Sqn
800B Sqn (refuel)
736 Sqn
764B Sqn
771 Sqn
776 FRU

31. de Havilland (HS) Sea Vixen
700Y Sqn
890 Sqn
892 Sqn
893 Sqn
899 Sqn
766 Sqn

32. Blackburn (HS) Buccaneer
700Z Sqn
800 Sqn
801 Sqn
803 Sqn
809 Sqn
700B Sqn
736 Sqn

33. British Aerospace Sea Harrier
700A Sqn
800 Sqn
801 Sqn
899 Sqn

Index

Aircraft and engines are indexed under name of manufacturer; numbered Service units referred to in the text, eg, squadrons and OCUs, are indexed under Royal Air Force and Royal Navy as appropriate.